Stephen Holler (Ed.)

Label-Free Sensing

MDPI

This book is a reprint of the Special Issue that appeared in the online, open access journal, *Sensors* (ISSN 1424-8220) in 2015 (available at: http://www.mdpi.com/journal/sensors/special_issues/label_free_sensing).

Guest Editor
Stephen Holler
Fordham Univeristy
USA

Editorial Office
MDPI AG
Klybeckstrasse 64
Basel, Switzerland

Publisher
Shu-Kun Lin

Managing Editors
Lin Li
Limei Huang

1. Edition 2016

MDPI • Basel • Beijing • Wuhan • Barcelona

ISBN 978-3-03842-210-5 (Hbk)
ISBN 978-3-03842-211-2 (PDF)

Table of Contents

V

List of Contributors

Ghasem Amoabediny Department of Biotechnology and Pharmacy Engineering, Faculty of Chemical Engineering, University of Tehran, Tehran 4563-11155, Iran; Research Center of New Technologies in Life Science Engineering, University of Tehran, Tehran 1417963891, Iran.

Mark Anderson Department of Biochemistry, University of Missouri, Columbia, MO 65201, USA.

Mariana Arruda Laboratório de Imunopatologia Keizo Asami (LIKA), Universidade Federal de Pernambuco-UFPE, Av. Prof. Moraes Rego, s/n, Campus da UFPE, 50670-901 Recife, PE, Brazil.

Carlota Bardina AntibodyBcn, MRB 104 Modul b UAB Campus, 08193 Bellaterra, Barcelona, Spain.

Matthew T. Bernards Department of Chemical Engineering, University of Missouri, Columbia, MO 65201, USA.

Wessulla Bezerra Laboratório de Imunopatologia Keizo Asami (LIKA), Universidade Federal de Pernambuco-UFPE, Av. Prof. Moraes Rego, s/n, Campus da UFPE, 50670-901 Recife, PE, Brazil.

Amélia Borba Laboratório de Imunopatologia Keizo Asami (LIKA), Universidade Federal de Pernambuco-UFPE, Av. Prof. Moraes Rego, s/n, Campus da UFPE, 50670-901 Recife, PE, Brazil.

Rabah Boukherroub Institute of Electronics, Microelectronics and Nanotechnology (IEMN), UMR-CNRS 8520, Université Lille 1, Avenue Poincaré—BP 60069, 59655 Villeneuve d'Ascq, France.

Baiqiong Cao Department of Electrical Engineering, Henan Agricultural University, Zhengzhou 450002, China.

Rafael Casquel Center for Biomedical Technology, Optics, Photonics and Biophotonics Lab, Universidad Politécnica de Madrid. Campus Montegancedo, 28223 Pozuelo de Alarcón, Madrid, Spain; Department of Applied Physics and Material, Escuela Técnica Superior de Ingenieros Industriales (ETSII), Universidad Politécnica de Madrid, Jose Gutiérrez Abascal, 2. 28006 Madrid, Spain.

Geunhyoung Chae College of Information and Communication Engineering, Sungkyunkwan University, Suwon 440-476, Korea.

Eric Chainet Laboratoire d'Electrochimie et de Physico-chimie des Matériaux et des Interfaces (LEPMI), 1130 rue de la Piscine BP75, 38402 Saint Martin d'Hères Cedex, France.

Keke Chang Department of Electrical Engineering, Henan Agricultural University, Zhengzhou 450002, China.

Ruipeng Chen Department of Electrical Engineering, Henan Agricultural University, Zhengzhou 450002, China.

Yannick Coffinier Institute of Electronics, Microelectronics and Nanotechnology (IEMN), UMR-CNRS 8520, Université Lille 1, Avenue Poincaré—BP 60069, 59655 Villeneuve d'Ascq, France.

Marli Cordeiro Departamento de Bioquímica, Universidade Federal de Pernambuco-UFPE, Av. Professor Moraes Rego, s/n, Campus da UFPE, CEP: 50670-901 Recife, PE, Brazil.

Brian T. Cunningham Department of Bioengineering, University of Illinois at Urbana-Champaign, Champaign, IL 61822, USA; Department of Electrical and Computer Engineering, University of Illinois at Urbana-Champaign, Champaign, IL 61822, USA.

Danielly Ferreira Laboratório de Imunopatologia Keizo Asami (LIKA), Universidade Federal de Pernambuco-UFPE, Av. Prof. Moraes Rego, s/n, Campus da UFPE, 50670-901 Recife, PE, Brazil.

Ebrahim Ghafar-Zadeh Department of Electrical Engineering and Computer Science, York University, Toronto, ON M3J1P3, Canada.

Ana L. Hernandez Center for Biomedical Technology, Optics, Photonics and Biophotonics Lab, Universidad Politécnica de Madrid. Campus Montegancedo, 28223 Pozuelo de Alarcón, Madrid, Spain.

Miguel Holgado Department of Applied Physics and Material, Escuela Técnica Superior de Ingenieros Industriales (ETSII); Center for Biomedical Technology, Optics, Photonics and Biophotonics Lab, Universidad Politécnica de Madrid. Campus Montegancedo, 28223 Pozuelo de Alarcón, Madrid, Spain.

Morteza Hosseini Department of Nanobiotechnology, School of New Sciences and Technologies, University of Tehran, Tehran 14395-1561, Iran.

Xinran Hu School of Human Nutrition and Dietetics, McGill University, Ste Anne de Bellevue, QC H9X 3V9, Canada.

Jiandong Hu Department of Electrical Engineering, Henan Agricultural University, Zhengzhou 450002, China; State key laboratory of wheat and maize crop science, Zhengzhou 450002, China.

Heather K. Hunt Department of Bioengineering, University of Missouri, Columbia, MO 65201, USA.

Daeho Jang School of Mechanical Engineering, Korea University, Seoul 136-701, Korea.

Mónica Jara AntibodyBcn, MRB 104 Modul b UAB Campus, 08193 Bellaterra, Barcelona, Spain.

Min Jiang College of life sciences, Henan Agricultural University, Zhengzhou 450002, China.

Carmen Jimenez Université Grenoble-Alpes, CNRS, Laboratoire des Matériaux et du Génie Physique (LMGP), MINATEC, 3 parvis Louis Néel, 38016 Grenoble Cedex 1, France.

Maríafe Laguna Department of Applied Physics and Material, Escuela Técnica Superior de Ingenieros Industriales (ETSII); Center for Biomedical Technology, Optics, Photonics and Biophotonics Lab, Universidad Politécnica de Madrid. Campus Montegancedo, 28223 Pozuelo de Alarcón, Madrid, Spain.

Alvaro Lavín Department of Applied Physics and Material, Escuela Técnica Superior de Ingenieros Industriales (ETSII); Center for Biomedical Technology, Optics, Photonics and Biophotonics Lab, Universidad Politécnica de Madrid. Campus Montegancedo, 28223 Pozuelo de Alarcón, Madrid, Spain.

Minh Hai Le Université Grenoble-Alpes, CNRS, Laboratoire des Matériaux et du Génie Physique (LMGP), MINATEC, 3 parvis Louis Néel, 38016 Grenoble Cedex 1, France; School of Materials Science and Engineering, Hanoi University of Science and Technology, 1 Dai Co Viet Street, 10000 Hanoi, Vietnam.

Jianwei Li Department of Electrical Engineering, Henan Agricultural University, Zhengzhou 450002, China.

Hao Liang Department of Electronic and Telecommunications, University of Gävle, Gävle SE-801 76, Sweden.

José Lima-Filho Laboratório de Imunopatologia Keizo Asami (LIKA), Universidade Federal de Pernambuco-UFPE, Av. Prof. Moraes Rego, s/n, Campus da UFPE, 50670-901 Recife, PE, Brazil; Departamento de Bioquímica, Universidade Federal de Pernambuco-UFPE, Av. Professor Moraes Rego, s/n, Campus da UFPE, CEP: 50670-901 Recife, PE, Brazil.

Kennya Lopes Departamento de Virologia e Terapia Experimental (LAVITE), Centro de Pesquisas Aggeu Magalhães (CPqAM), Fundação Oswaldo Cruz (Fiocruz)—Pernambuco, Av. Professor Moraes Rego, s/n, Campus da UFPE, 50.670-420 Recife, PE, Brazil.

Liuzheng Ma Department of Electrical Engineering, Henan Agricultural University, Zhengzhou 450002, China.

Danyelly Martins Laboratório de Imunopatologia Keizo Asami (LIKA), Universidade Federal de Pernambuco-UFPE, Av. Prof. Moraes Rego, s/n, Campus da UFPE, 50670-901 Recife, PE, Brazil; Departamento de Bioquímica, Universidade Federal de Pernambuco-UFPE, Av. Professor Moraes Rego, s/n, Campus da UFPE, CEP: 50670-901 Recife, PE, Brazil.

Gustavo Nascimento Laboratório de Imunopatologia Keizo Asami (LIKA), Universidade Federal de Pernambuco-UFPE, Av. Prof. Moraes Rego, s/n, Campus da UFPE, 50670-901 Recife, PE, Brazil.

Natália Oliveira Laboratório de Imunopatologia Keizo Asami (LIKA), Universidade Federal de Pernambuco-UFPE, Av. Prof. Moraes Rego, s/n, Campus da UFPE, 50670-901 Recife, PE, Brazil.

Yong-le Pan U.S. Army Research Laboratory, 2800 Powder Mill Road, Adelphi, MD 20783, USA.

Mojgan Ahmadzadeh Raji Department of Nanobiotechnology, School of New Sciences and Technologies, University of Tehran, Tehran 14395-1561, Iran; Research Center of New Technologies in Life Science Engineering, University of Tehran, Tehran 1417963891, Iran.

Brandon Redding U.S. Army Research Laboratory, 2800 Powder Mill Road, Adelphi, MD 20783, USA.

Beatriz Santamaría Center for Biomedical Technology, Optics, Photonics and Biophotonics Lab, Universidad Politécnica de Madrid. Campus Montegancedo, 28223 Pozuelo de Alarcón, Madrid, Spain; BioOpticalDetection, Centro de Empresas de la UPM, Campus Montegancedo, 28223 Pozuelo de Alarcón, Madrid, Spain.

Francisco J. Sanza Center for Biomedical Technology, Optics, Photonics and Biophotonics Lab, Universidad Politécnica de Madrid. Campus Montegancedo, 28223 Pozuelo de Alarcón, Madrid, Spain; BioOpticalDetection, Centro de Empresas de la UPM, Campus Montegancedo, 28223 Pozuelo de Alarcón, Madrid, Spain.

Mark J. Schwab U.S. Army Research Laboratory, 2800 Powder Mill Road, Adelphi, MD 20783, USA.

Sehyun Shin School of Mechanical Engineering, Korea University, Seoul 136-701, Korea.

Javier Soria Bioftalmik. Parque Tecnológico Zamudio Ed. 800 2ª Planta 48160, Bizkaia, Spain.

Elaine Souza Universidade Federal de Alagoas (UFAL), Campus Arapiraca, Av. Manoel Severino Barbosa, s/n, Bom Sucesso, 57.309-005 Arapiraca, AL, Brazil.

Valerie Stambouli Université Grenoble-Alpes, CNRS, Laboratoire des Matériaux et du Génie Physique (LMGP), MINATEC, 3 parvis Louis Néel, 38016 Grenoble Cedex 1, France.

Tatiana Suarez Bioftalmik. Parque Tecnológico Zamudio Ed. 800 2ª Planta 48160, Bizkaia, Spain.

Xiaohui Sun Department of Electrical Engineering, Henan Agricultural University, Zhengzhou 450002, China.

Sabine Szunerits Institute of Electronics, Microelectronics and Nanotechnology (IEMN), UMR-CNRS 8520, Université Lille 1, Avenue Poincaré—BP 60069, 59655 Villeneuve d'Ascq, France.

Parviz Tajik Department of Theriogenology, Faculty of Veterinary Medicine, University of Tehran, Tehran 1419963111, Iran.

Fanyongjing Wang Department of Bioengineering, University of Missouri, Columbia, MO 65201, USA.

Shun Wang Department of Electrical Engineering, Henan Agricultural University, Zhengzhou 450002, China.

Deborah Zanforlin Laboratório de Imunopatologia Keizo Asami (LIKA), Universidade Federal de Pernambuco-UFPE, Av. Prof. Moraes Rego, s/n, Campus da UFPE, 50670-901 Recife, PE, Brazil.

Juanhua Zhu Department of Electrical Engineering, Henan Agricultural University, Zhengzhou 450002, China.

Yue Zhuo Department of Bioengineering, University of Illinois at Urbana-Champaign, Champaign, IL 61822, USA.

About the Guest Editor

Stephen Holler received his PhD in applied physics from Yale University studying means for remotely characterizing airborne particles through light scattering and fluorescence spectroscopy. After graduating, he joined Los Gatos Research, a small R&D company in the San Francisco Bay Area, where he worked on laser-based diagnostics and ultra-sensitive detection techniques. After 9/11, Dr. Holler joined the Lasers, Optics and Remote Sensing group at Sandia National Laboratories in Albuquerque, NM focusing on optical techniques for biological particle detection. After a brief stay at Sandia, Dr. Holler joined NovaWave Technologies as Director of R&D. NovaWave focused on developing laser-based sensors for environmental monitoring. In 2010, NovaWave was acquired by Thermo Fisher Scientific. Dr. Holler remained with Thermo Fisher before joining the faculty at Fordham University in the Department of Physics and Engineering Physics in 2011. His work in the Laboratory on micro-optics and biophotonics employs optical microcavities to perform sensitive label-free detection of bionanoparticles, Raman spectroscopy of tissue samples for cancer diagnostics, and light scattering for aerosol particle studies. In addition, Dr. Holler oversees the Fordham Seismic Station and is involved in expanding 3D printing and robotics capabilities at Fordham.

Introduction to the Special Issue on Label-Free Sensing

Stephen Holler

The implementation of label-free sensing of biological and chemical agents allows one to investigate the underlying physical and chemical characteristics and interactions of target analytes while reducing both sample complexity and preparation time. Sensor platforms incorporating label-free detection schemes avoid the potentially confounding effects of molecular labels by monitoring the target species directly, relying solely on the intrinsic physicochemical properties of the target analyte. Because of the relatively minimal sample preparation, such approaches are well suited for field applications and remote diagnostics where either sample preparation facilities and/or trained personnel may be limited or unavailable. This special issue highlights some diverse approaches to the challenge of detecting target analytes without the need for labels. These approaches principally focus on optical and electrochemical techniques, and offer the promise of a rapid diagnostics tool that could be used in a clinical setting that would minimize the time between identification and treatment.

Reprinted from *Sensors*. Cite as: Holler, S. Introduction to the Special Issue on Label-Free Sensing. *Sensors* **2015**, *4*, 623–636.

"The single biggest threat to man's continued dominance on the planet is the virus." These ominous words belong to Nobel Laureate Joshua Lederberg, and while he believed that virus poses an existential threat to humanity, mankind faces a litany of attacks from no less deadly threats, both naturally occurring and man-made. In order to effectively combat this onslaught it is vital that one be able to effectively identify the threat, as identification is the first step in the treatment. Treatment is crucial because without it maintenance of health, and protection from chemical and biological threats would be impossible. Sensitive instrumentation is needed to initially identify a threat in order to diagnosis a disease or negative impact of exposure, but sensors also play an important role in providing some quantifiable metric by which post-treatment efficacy can be gauged.

A host of methodologies exist for identifying and characterizing threats, both known and unknown. Technologically derived methods permit an enhanced sensor response by incorporating labels, probes that bind to the target analyte to improve detection capability. Often these are fluorophores that provide an indirect means for sensing the presence of some species. The use of such probes is widespread and can facilitate ultrasensitive detection by boosting the signal-to-noise ratio of a measurement. However, there are disadvantages to the use of probes. For example, affixing probes to malignant tissue will cause cancerous cells to shine brightly, but this can obscure tumor margins. Furthermore, in these situations the probes only provide surface coverage and yield no information about the depth of the malignancy. *In vivo* studies using probes can be delicate since many highly effective probes have inherent toxicity to humans. Quantum dots are a prime example; they offer great promise for labeling in *ex vivo* analysis but are unsuitable for injection into patients, and have fallen by the wayside in this regard. Since disease detection and treatment will be the greatest threat that we face it is crucial that any sensing technology for *in vivo* applications employ low-toxicity biocompatible materials. In this sense, the ideal sensing modality would be based on the inherent properties of the target species. The ideal sensor would be label-free.

The implementation of label-free techniques for sensing biological and chemical agents has grown considerably in recent years. New approaches are being developed that allow one to investigate the underlying physical and chemical characteristics and interactions of target analytes while reducing both sample complexity and preparation time. In addition, these sensor platforms avoid the potentially confounding effects and potentially hazardous effects of molecular labels by monitoring the target species directly, relying solely on the target's intrinsic physicochemical properties. Because of the relatively minimal sample preparation, such approaches are well suited for field applications and remote diagnostics where either sample preparation facilities and/or trained personnel may be limited or unavailable.

This special issue is devoted to label-free sensing techniques that may be used in a wide variety of applications from biodefense to cancer screening to mass spectrometry. This compilation is by no means complete, but it does provide a good survey of techniques that researchers are using to perform label-free sensing. There are both original contributions and review articles that summarize the state-of-the-art. This issue is loosely divided into two sections that broadly categorize these contributions to the label-free sensing literature: optical and electrochemical. Since both of these categories are broad there is some overlap in the work they encompass, however they generally cover a number of different techniques that have been demonstrated to effectively perform the task at hand.

Immediately what comes to mind for optical approaches are spectroscopic techniques. The use of spectroscopy for characterizing samples is venerable, in part because the molecular constituents of matter interact with electromagnetic radiation and elicit a response. These interactions are, after all, the basis for vision, the most universal label-free sensing mechanism. However, enhancements in detection capability may be made by incorporating new sensor morphologies or new optical materials. Consequently, improvements to signal-to-noise may be achieved, and ever decreasing detection limits may be observed, with the ultimate goal being single molecule detection.

Electrochemical sensing modalities are another natural progression in the development of label-free sensing devices. Again, our basic operation is governed by electrochemical interactions; the heart would not beat and the brain would cease to function if their intrinsic electrical properties were eliminated. Despite the heart and brain both having underlying electrochemical properties, their composition is dramatically different. It is the unique response of the cellular components that allow electrochemical interactions to provide sensing discrimination. Furthermore, electrical and chemical measurements also have a long history, and the continued improvement of materials and high precision/high sensitivity instrumentation is allowing researchers to gain better understanding of the physical and chemical responses of target analytes.

Fundamental to optical spectroscopy is the manner in which molecules move. Whether it is through rotations, vibrations, or some combination thereof, molecules leave their fingerprints on electromagnetic radiation. This present compilation begins with a review of Raman spectroscopy on isolated bioaerosols from researchers at the Army Research Lab and Yale University [1]. The ability to isolate and suspend a particle frees it from interfering effects associated with containment vessels, leaving only the signal from the aerosol. These signals are species specific and may be used for discrimination and classification. Complete characterization of bioaerosols remains a challenge, but is crucial to maintaining a healthy environment and addressing the threat of bioterrorism.

Microscopy is a venerable technique for studying microscopic entities. However, spatial discrimination, particularly for small molecules can be challenging. Fluorescence microscopy can be used to improve detection capabilities. Researchers at University of Illinois at Urbana-Champaign review the state of photonic crystal enhanced microscopy [2]. Photonic crystals are used to manipulate the optical characteristics of a material through nanostructured surfaces. Optical enhancements provide a sensitive means for detecting broad classes of materials such as dielectric nanoparticles, plasmonic nanoparticles, biomolecular layers, and cells. These broad capabilities allow researchers to examine a host of processes, with the ultimate goal of achieving single molecule detection resolution.

Surface plasmon resonance (SPR) offers a sensitive means for detecting trace species of a target analyte. The plasmon resonance boosts electric field strength locally leading to improved detection capabilities. Often detection capabilities are hindered by the ability to appropriately fit changes in the measured signal, especially when fits to nonlinear curves are based on simple polynomial regressions. Researchers at Korea University and Sungkrunkwan University have tackled this problem by developing a new sigmoid-asymmetric fitting routine [3].

The results are in excellent agreement over the full SPR curve, which leads to improved resolution and detection sensitivity. While a collaborative effort among Henan Agricultural University, McGill University, and University of Gälve has sought to improve SPR with high performance A/D and custom signal amplifiers [4]. The goal of this work, like many sensor projects, is the development of a compact, low-cost, fieldable instrument. Presently in the laboratory stage, the compact sensor has demonstrated good detection capabilities and is being prepared for field work.

Enzyme-linked immunosorbent Assay (ELISA) offers a high standard for detection, but it requires the use of labels. The development of a competitive approach that is label-free would be a boon to researchers and clinical diagnosticians. Work out of Universidad Politécnica de Madrid has demonstrated just this [5]. Using Fourier Transform Visible-Infrared Spectrometry coupled with a Fabry-Pérot inteferometer they were able to develop an immunoassay approach with response comparable to ELISA, but label-free. Specifically they targeted biomarkers associated with dry eye dysfunction.

Whispering gallery mode biosensors have emerged in the last fifteen years as powerful tools in ultrasensitive detection. They have been used to demonstrate detection of DNA hybridization, bacteria, virus, and even single protein molecules. However, in mixed media these, like many other sensor platforms, are subject to non-specific adsorption. Research out of the Department of Bioengineering at the University of Missouri seeks to minimize the confounding effects of nonspecific adsorption using poly(ethylene glycol) to form a nonfouling surface layer in conjunction with specific biorecognition elements [6]. This is especially important to minimize scavenging and non-efficient binding to regions outside the sensing mode volume.

Carbon nanotubes and graphene have emerged as key components in an array of mechanical and electrochemical sensing applications. However, less well-known alternatives such as diamond nanowires offer a fertile platform for researchers. Due to their inherently advantageous properties such as biocompatibility, chemical inertness, high conductivity (electrical and thermal), and high mechanical strength. Researchers at the Institute of Electronics at the Université Lille 1 are leveraging the properties of diamond nanowires, specifically boron-doped diamond nanowires, to develop novel platforms for electrochemistry and mass spectrometry [7]. The ultimate goal being to combine the electrochemical sensing approach with the mass spectrometry to create a platform for electrochemically enhanced mass spectrometry which would benefit researchers in a number of different fields.

Impedance sensors offer a platform to detect a wide range of substances. These sensors work on a number of vapor phase targets to detect a host of environmental hazards. A collaborative effort between the Université Grenoble-

Alpes and Hanoi University of Science and Technology has taken these platforms to the next level. Using nanoporous SnO_2 they have developed a label-free impedimetric sensing platform [8]. Their device demonstrates detection capabilities in both the liquid and vapor phases while offering discrimination capabilities down to a single base mismatch in DNA studies. The high sensitivity and selectivity with a label-free platform enables a host of DNA hybridization experiments to be performed.

The final two papers of this compilation tackle real diseases that affect millions of people globally: Dengue Virus [9] and Colon Cancer [10]. The work on the detection of the Dengue virus comes from the Universidade Federal de Pernambuco-UFPE, Universidade Federal Alagoas, and Centro de Pesquisas Aggeu Magalhães. This collaborative effort utilizes pencil graphite electrodes to perform differential pulse voltammetry to characterize the response of sequences of Dengue Serotype 3. They achieved high sensitivity and selectivity in a platform that has the potential to be both a fast and inexpensive method for serotype identification. The colon cancer work was performed jointly by researchers at the University of Tehran and York University, and employed aptamer functionalized electrodes for a battery of tests including flow cytometry, fluorescence microscopy, and electrochemical cyclic voltammetry. Their approach has demonstrated limits of detection of less than 10 cancer cells, which offers the promise for rapid point-of-care diagnostics.

The work presented in this special issue is a subset of the continually growing field of label-free sensing. The diversity offered by these papers exhibits just a fraction of the range of detection methodologies being pursued. These papers provide insight into the field and demonstrate that ultrasensitive detection is possible and may one day soon find its way into clinical facilities for rapid diagnostics thus reducing the time between identification and treatment. The best defense may be a good offense, and early detection enables implementation of the best offense one could hope for.

Acknowledgments: I wish to thank all the authors and the reviewers for all their contributions to this body of work.

Conflict of Interests: There are no conflicts of interested associated with this paper.

References

1. Redding, B.; Schwab, M.J.; Pan, Y.l. Raman Spectroscopy of Optically Trapped Single Biological Micro-Particles. *Sensors* **2015**, *15*, 19021.
2. Zhuo, Y.; Cunningham, B.T. Label-Free Biosensor Imaging on Photonic Crystal Surfaces. *Sensors* **2015**, *15*, 21613.
3. Jang, D.; Chae, G.; Shin, S. Analysis of Surface Plasmon Resonance Curves with a Novel Sigmoid-Asymmetric Fitting Algorithm. *Sensors* **2015**, *15*, 25385.

4. Chang, K.; Chen, R.; Wang, S.; Li, J.; Hu, X.; Liang, H.; Cao, B.; Sun, X.; Ma, L.; Zhu, J.; Jiang, M.; Hu, J. Considerations on Circuit Design and Data Acquisition of a Portable Surface Plasmon Resonance Biosensing System. *Sensors* **2015**, *15*, 20511.

5. Laguna, M.; Holgado, M.; Hernandez, A.L.; Santamaría, B.; Lavín, A.; Soria, J.; Suarez, T.; Bardina, C.; Jara, M.; Sanza, F.J.; Casquel, R. Antigen-Antibody Affinity for Dry Eye Biomarkers by Label Free Biosensing. Comparison with the ELISA Technique. *Sensors* **2015**, *15*, 19819.

6. Wang, F.; Anderson, M.; Bernards, M.T.; Hunt, H.K. PEG Functionalization of Whispering Gallery Mode Optical Microresonator Biosensors to Minimize Non-Specific Adsorption during Targeted, Label-Free Sensing. *Sensors* **2015**, *15*, 18040.

7. Szunerits, S.; Coffinier, Y.; Boukherroub, R. Diamond Nanowires: A Novel Platform for Electrochemistry and Matrix-Free Mass Spectrometry. *Sensors* **2015**, *15*, 12573.

8. Le, M.H.; Jimenez, C.; Chainet, E.; Stambouli, V. A Label-Free Impedimetric DNA Sensor Based on a Nanoporous SnO_2 Film: Fabrication and Detection Performance. *Sensors* **2015**, *15*, 10686.

9. Oliveira, N.; Souza, E.; Ferreira, D.; Zanforlin, D.; Bezerra, W.; Borba, M.A.; Arruda, M.; Lopes, K.; Nascimento, G.; Martins, D.; Cordeiro, M.; Lima-Filho, J. A Sensitive and Selective Label-Free Electrochemical DNA Biosensor for the Detection of Specific Dengue Virus Serotype 3 Sequences. *Sensors* **2015**, *15*, 15562.

10. Raji, M.A.; Amoabediny, G.; Tajik, P.; Hosseini, M.; Ghafar-Zadeh, E. An Apta-Biosensor for Colon Cancer Diagnostics. *Sensors* **2015**, *15*, 22291.

Raman Spectroscopy of Optically Trapped Single Biological Micro-Particles

Brandon Redding, Mark J. Schwab and Yong-le Pan

Abstract: The combination of optical trapping with Raman spectroscopy provides a powerful method for the study, characterization, and identification of biological micro-particles. In essence, optical trapping helps to overcome the limitation imposed by the relative inefficiency of the Raman scattering process. This allows Raman spectroscopy to be applied to individual biological particles in air and in liquid, providing the potential for particle identification with high specificity, longitudinal studies of changes in particle composition, and characterization of the heterogeneity of individual particles in a population. In this review, we introduce the techniques used to integrate Raman spectroscopy with optical trapping in order to study individual biological particles in liquid and air. We then provide an overview of some of the most promising applications of this technique, highlighting the unique types of measurements enabled by the combination of Raman spectroscopy with optical trapping. Finally, we present a brief discussion of future research directions in the field.

Reprinted from *Sensors*. Cite as: Redding, B.; Schwab, M.J.; Pan, Y. Raman Spectroscopy of Optically Trapped Single Biological Micro-Particles. *Sensors* **2015**, *15*, 19021–19046.

1. Introduction

Raman spectroscopy relies on measuring the frequency and relative intensity of inelastically scattered light due to the vibrational, rotational, and other low-frequency modes of a sample. As such, the Raman spectrum provides a fingerprint of the molecules present in a sample [1,2]. Raman spectroscopy has been broadly used as one of the main diagnostic techniques in analytical chemistry and is developing into an important method in biology and medicine as a real-time clinical diagnostic tool for the identification of disease, and evaluation of living cells and tissue [1]. In addition, Raman spectroscopy is a promising method for the identification of aerosolized biological and chemical threat agents.

The primary challenge associated with performing Raman spectroscopy is the inefficiency of the Raman scattering process, which results in a signal ~100 dB weaker than typical fluorescence [2]. Hence, spontaneous Raman measurements require a long signal integration time and can be difficult to perform on individual cells or particles in a solution or in the air which do not remain in the same position long enough to acquire a Raman spectrum. One solution to this challenge is to deposit the particle or cell of interest on a substrate before the measurement [3]. However, this has clear limitations since the substrate can alter the Raman spectrum of the particle, limit the ability to perform longitudinal studies of a particle in its natural environment, or introduce a background Raman signal, making it difficult to isolate the Raman spectrum from the particle of interest [4]. Moreover, dense particle deposition introduces challenges when trying to obtain the Raman spectrum from a single particle.

The combination of laser trapping with Raman spectroscopy (LTRS) circumvents these issues by holding a particle or cell in place long enough for data acquisition. Since optical trapping is possible in both solution and air, the potential influence of inelastic scattering from the substrate is avoided [4]. Since the Raman spectrum of a trapped particle can be measured *in situ*, studies on the temporal response of a particle to environmental changes are possible [5]. In addition, particle trapping using laser tweezers holds the particle near the high intensity portion of the beam, simplifying the alignment by maximizing the Raman signal. Such a combined method also enables the study of individual particles, providing information about the heterogeneity of a population which can be difficult to extract from a Raman measurement of a bulk sample [6]. Performing LTRS on relatively large biological particles can even enable the measurement of the molecular content of different regions of a cell [7].

While LTRS has been performed on a wide range of particle types, in this review we will focus on its application to the characterization of biological particles. Biological aerosol particles, or bioaerosols, have important implications for human health, acting as airborne disease transmitters that contain microorganisms such as bacteria, viruses, pollen, and fungi. Monitoring bioaerosols in locations such as hospitals for the presence of airborne diseases, or public spaces for the detection of aerosolized biological warfare agents are increasingly important problems. Aerosol particles also have significant implications for climate change due to their role in the scattering and absorption of solar radiation as well as in cloud condensation and the formation of ice nuclei. Thorough characterization of the composition and density of aerosol particles is therefore essential to the accuracy of climate change models. Raman spectroscopy, particularly when combined with optical trapping, is uniquely suited to the characterization of bioaerosols due to its combination of high specificity with a modest cost and non-invasive nature.

LTRS is also emerging as a powerful tool in molecular biology due to its ability to perform longitudinal studies on individual cells, spores, bacteria, and viruses in their natural environments. Bioaerosols are a complex mixture containing numerous biomolecules in various concentrations and forms. Previous single-particle optical characterizations using fluorescence were only able to probe a limited range of biological compounds, including proteins, amino acids (tyrosine, tryptophan *etc.*), nucleic acids (DNA, RNA *etc.*), coenzymes (nicotinamide adenine dinucleotides, flavins, and vitamins B_6 and K and variants of these), polysaccharides, dipicolinates, and lipids. Raman spectroscopy, especially when long acquisition times are enabled through LTRS, can characterize a much broader range of biomolecules and with higher specificity compared to techniques that probe only fluorescent compounds. Some cells or spores can grow, change, and reproduce in buffer liquid or in air, and LTRS enables the study of these cells as they undergo these processes. For example, as a cell grows, some biomolecules can decrease or vanish, while others increase or can even be generated. Therefore, using Raman spectroscopy to detect and monitor specific biomolecules within a cell as it responds to changes in its environment can provide new insights into our fundamental understanding of cell growth. LTRS is also emerging as an important tool in drug discovery due to its ability to monitor a cells response, for example, to varying forms of chemotherapy [8].

In this paper, we provide a brief review of techniques that perform Raman spectroscopy on individual optically trapped biological particles. We discuss many of the promising applications of LTRS and attempt to highlight the unique features of LTRS which make it such a powerful technique. This paper is organized as follows: in Section 2, we present a discussion of the most common optical trapping techniques used in LTRS; in Section 3 we discuss the development of LTRS as well as several exemplary applications. In Section 4, we provide a brief overview and discussion of future applications and research directions. We hope this review will provide researchers entering the field with an introduction to the wide array of LTRS applications as well as its key features.

2. Optical Trapping Techniques

Optical tweezers rely on the radiative pressure force which results from the transfer of momentum from photons to a particle. In Figure 1a, we illustrate the influence of the radiative pressure force on a particle in a collimated beam and in a focused beam. The radiative pressure force is often divided into a scattering force and a gradient force, although both result from the same transfer of momentum from the incident photons [9,10]. The scattering force tends to push particles in the direction of light propagation whereas the gradient force tends to pull the particle towards the high intensity region. Ashkin's original demonstration relied on the radiative pressure acting on "relatively transparent particles in a relatively transparent media" to avoid thermal effects which "are usually orders of magnitude larger than radiation pressure" for strongly absorbing particles [11]. Absorbing particles are subject to a photophoretic force which results when an absorbing particle is non-uniformly heated and/or non-uniformly heat-emitting. As illustrated in Figure 1b, a strongly absorbing particle is non-uniformly heated if it is illuminated from one side. When the heat is transferred to the surrounding gas molecules, gas molecules on the warmer side of the particle will acquire more energy and subsequently collide with the particle at higher velocity, imposing a net force pushing the particle toward its cold side. This photophoretic force can be 4–5 orders of magnitude stronger than the radiative pressure force [12] and is therefore the dominant force acting on strongly absorbing particles.

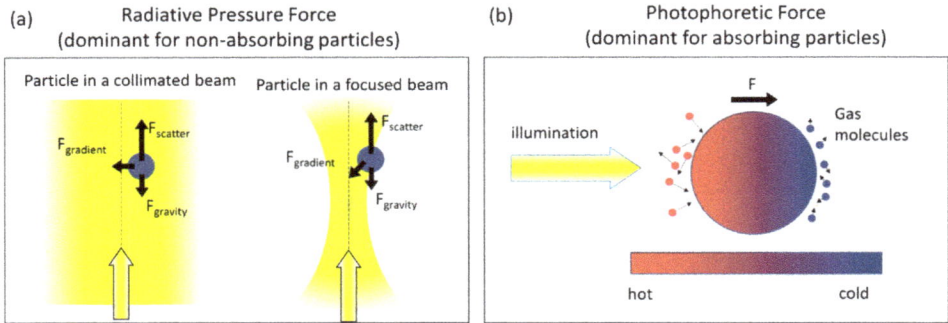

Figure 1. (**a**) The radiative pressure force, which is the dominant force experienced by non-absorbing particles, results from the transfer of momentum from photons scattered by a particle. The radiative pressure force can be divided into a scattering force, which tends to push the particle along the direction of light propagation, and a gradient force, which tends to pull the particle toward the highest intensity region. The gradient force enables trapping in a focused laser beam; (**b**) The photophoretic force, which is the dominant force experienced by strongly absorbing particles, results from the transfer of heat to surrounding gas molecules from a non-uniformly heated and/or non-uniformly heat-emitting particle.

2.1. Optical Trapping via the Radiative Pressure Force (Laser Tweezers)

In 1970, Ashkin first demonstrated that optical radiation pressure could be used to trap glass beads in water using two counter propagating, focused beams [11]. A year later, he demonstrated levitation of glass spheres in air and in vacuum using a vertically oriented beam to compensate gravity [13], and in 1976, Roosen *et al.* showed that the gradient force was strong enough to overcome gravity, enabling the trapping of solid glass spheres in two counter propagating horizontal beams [14]. The first single beam optical trap for an airborne particle (a 5 μm glass sphere) was demonstrated in 1997 by using an objective with a high numerical aperture (NA = 0.95) to provide a sufficiently strong gradient force [15]. Since these initial demonstrations the field of optical tweezers has experienced rapid growth and developed into an indispensable tool in the study and manipulation of micron sized particles [16].

Radiative pressure based optical trapping techniques can be divided into single or multiple beam configurations. Single beam traps are more easily aligned; however, a high NA is typically required to enable optical trapping. This constraint is particularly pronounced when trapping particles in air, since the high refractive index contrast between the particle and air results in a strong scattering force which tends to destabilize the trap [9,17]. Using two counter-propagating beams to cancel out the scattering force enables optical trapping of airborne particles with much lower NA (Figure 2); however the alignment in such systems can be very critical [9].

Radiative pressure traps have been demonstrated with both continuous wave (CW) and pulsed lasers. Although the average power was found to be the primary factor dictating the efficacy of the optical trap [18], trapping using a pulsed laser may have advantages in potential non-linear optical applications.

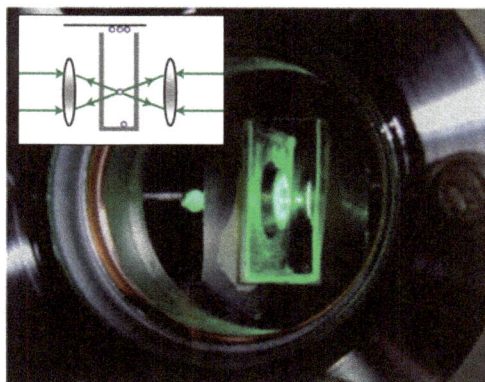

Figure 2. A 4.7 μm diameter microsphere trapped inside a vacuum chamber by a counter-propagating dual-beam optical tweezer. The wavelength of the trapping beams is 1064 nm; A weak green (532 nm) laser is used for illumination. Inset is a counter-propagating dual-beam optical trap in air based on radiative pressure forces. With kind permission from Springer Science and Business Media [9].

2.2. Optical Trapping via the Photophoretic Force

The photophoretic force can provide a highly stable optical trap even for airborne particles. Optical levitation based on the photophoretic force was demonstrated as early as 1982 [12] and photophoretic trapping in a low-light optical vortex in 1996 [19]. In recent years, a number of additional techniques have been developed which utilize the photophoretic force to trap airborne particles. Unlike laser tweezers, optical traps based on the photophoretic force generally trap absorbing particles in a low-light intensity region where the particle is surrounded by light in 3-dimensions, as in the example shown in Figure 3 where a particle is trapped between two counter-propagating vortex beams [20–22]. Additional methods to generate such a low-light intensity region include hollow cones formed by a ring illuminating the back aperture of a lens [23,24], a low-light region formed between two counter-propagating hollow beam [24], tapered rings [25], optical lattices [26], bottle beams [27], and even speckle fields [28]. Although absorbing particles were trapped in the low-light region in each of these demonstrations, there have also been a few recent demonstrations of optical trapping in the high-intensity portion of a single focused beam [29,30]. To explain the origin of this phenomena, researchers have cited the role of the accommodation coefficient, which describes the ability of a particle to transfer heat to the surrounding gas molecules [31–33]. The accommodation coefficient depends on the material and morphology of a particle. If the accommodation coefficient varies along the surface of a particle, a body-centric force can result even in a uniformly heated particle. Moreover, the accommodation force can at times be orders of magnitude stronger than the "longitudinal" photophoretic force (*i.e.*, the force shown in Figure 1b) [32], and could explain anomalous observations such as a "negative" photophoretic force experienced by strongly absorbing particles [34,35].

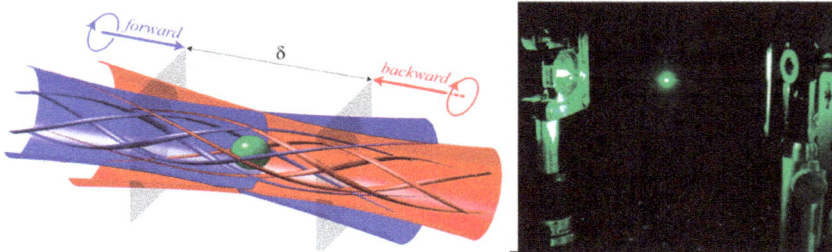

Figure 3. An example of a photophoretic trap. The particle is trapped in the low intensity region between two counter-propagating Laguerre-Gaussian vortex beams [21] (Fair Use according to OSA).

2.3. Alternate Trapping Modalities

Holographic optical tweezers enables many particles to be trapped and manipulated simultaneously. It was first demonstrated using a fixed diffractive optical element [36–38]. However, the functionality of holographic optical tweezers was greatly enhanced by the development of spatial light modulators which enabled researchers to rapidly update the optical trapping pattern without the need to fabricate a new diffractive optical element [38,39]. This approach was also applied to trap aerosol droplets [40].

Recently, optical fibers have also been proposed as a mechanism to achieve optical trapping. For example, Jess *et al.* [41] showed that a particle could be trapped in the diverging beams between two multimode fibers directed toward each other, as shown in Figure 4. This method enabled the manipulation of larger cells (up to 100 μm in diameter) than can be trapped in most optical tweezers systems [41]. A separate microscope objective was then used to collect the Raman spectra of the trapped particles, providing a means to collect Raman spectra from different positions within a trapped cell. Analysis of the spatially varying Raman spectra within the cell were used to allow for the identification of the nucleus, cytoplasm, and membrane regions of the cell using a principal component analysis (PCA) [41]. The dual fiber trap was also extended to trap and record the Raman spectra from particles in microfludic flow channels, as shown in Figure 4. This illustrates the potential for LTRS to be used for the on-line characterization of particles in a microfluidic system.

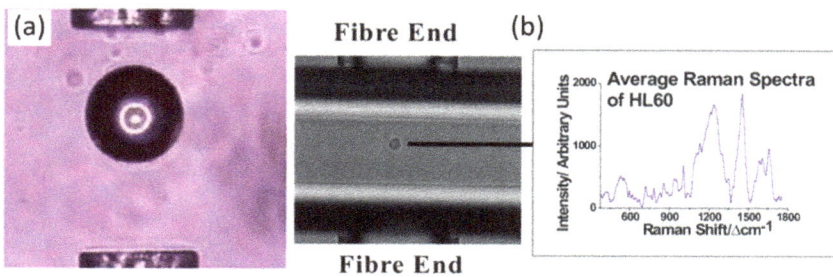

Figure 4. (**a**) A 100 μm polymer sphere is trapped between two fibers; (**b**) A HL60 cell is trapped in a microfluidic channel between two fibers. The particle is stopped in flow while the Raman spectrum is recorded and then released [41] (Fair Use according to OSA).

Optical trapping has also been demonstrated using an individual multimode fiber. In this configuration a spatial light modulator controlled the wavefront of light coupled into the fiber to form a focal spot (or several focal spots) at the distal end of the fiber to enable optical trapping. In this way, the system resembled holographic optical tweezers extended through a multimode fiber [42,43].

2.4. Trapping both Transparent and Absorbing Particles in Air Using a Single Shaped Laser Beam

Due to the distinct nature of the radiative pressure and photophoretic forces, most optical traps formed by a single laser beam are designed for either trapping absorbing or transparent particles. However, many applications require the ability to trap particles regardless of their morphology and absorptivity. Recently, a technique was shown to enable the trapping of both absorbing and transparent particles using a fixed optical geometry [44]. In this approach, a single shaped laser beam forms a hollow optical cone in which absorbing particles are trapped in the low-light-intensity region above the focal spot via the photophoretic force while non-absorbing particles are trapped at the high-intensity focal spot via the radiative pressure force. The experimental trapping apparatus used to realize this optical trap is shown in Figure 5a along with the calculated intensity profile near the focal spot (shown on a log-scale), an image of the conical focal region, and an image of a Johnson smut grass spore trapped in air [44]. This approach also reduces the scattering force near the focal spot, thereby enabling radiative pressure based trapping of transparent particles with lower NA (e.g., N ~ 0.55 for a particle with refractive index of 1.5) compared with traditional laser tweezers which require NA ~ 0.9 [17,44,45]. This approach was first used to trap droplets in air [45] and later shown to trap solid, transparent particles such as glass beads and albumin in air, as well as absorptive particles such as fungal spores [44]. Moreover, since particles of each type are trapped along the optical axis, this method could be combined with a particle interrogation technique such as Raman spectroscopy by imaging the optical axis to the entrance slit of a spectrometer. The ability to trap airborne particles in a fixed optical geometry regardless of the particle morphology or absorptivity could enable extensive on-line characterization of bioaerosols.

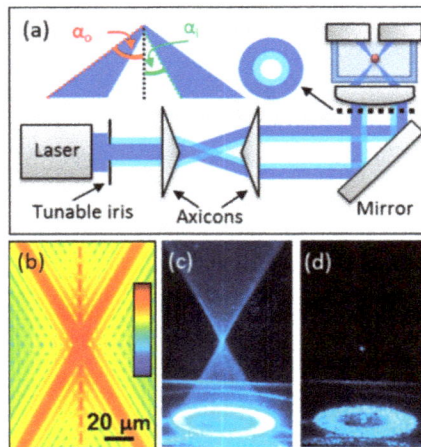

Figure 5. (a) Schematic of the optical trapping apparatus used to trap both transparent and absorbing airborne particles of arbitrary morphology using a single shaped hollow laser beam. The aspheric lens forms a hollow conical focus within a glass chamber where airborne particles are trapped; (b) Calculated intensity profile near the focal spot plotted on a log-scale; (c) Image of the conical focal region produced inside the chamber obtained by introducing Johnson Smut Grass Spores and recording a long exposure time image; (d) Image of a spore trapped in air near the focal point [44] (Fair Use according to OSA).

In addition to the optical techniques discussed above, bioaerosols particles can also be trapped using magnetic [46], electrodynamic [47], and acoustic forces [48]. However, in this article, we will limit our discussion to optical trapping techniques and their integration with Raman spectroscopy.

3. Laser Trapping Raman Spectroscopy (LTRS)

3.1. Development of LTRS

Raman spectroscopy was first combined with optical trapping in a 1984 work in which the Raman spectra were measured from levitated glass spheres and quartz microcrystals in air [4]. Soon after, optical trapping was used to obtain information about the molecular structure from single microdroplets [49,50]. Later, a near-infrared (NIR) laser source was shown to reduce the fluorescence background and photo-damage effects on live cells, although it increased the alignment complexity and the instrument cost [51,52]. The first study performed on biological particles was not conducted until 2002 when LTRS was demonstrated on single cellular organelles [53] as well as on living blood cells and yeast cells [54]. Soon afterwards, it was applied to obtain surface-enhanced Raman scattering (SERS) from single optically trapped bacterial spores [55]. As an example, Figure 6 shows the Raman spectra recorded from optically trapped yeast cells, illustrating the ability of Raman spectroscopy to differentiate between live and dead yeast cells which are essentially indistinguishable from the microscope image.

A 2003 study demonstrated the ability of LTRS to study the behavior of a single cell over time as it responded to environmental changes [5]. In particular, the response of single cells of *Escerichia coli* and *Enterobacter aerogenes* bacteria to changes in temperature was studied. The study observed significant changes in the phenylalanine band which was attributed to heat denaturation of proteins [5]. The temporal dynamics of yeast cells exposed to changing temperatures were studied via LTRS a year later [56]. Raman spectra of the trapped yeast cells showed irreversible changes in two of the Raman lines (1004 and 1604 cm^{-1}) as temperature increased from 25 °C to 80 °C [56].

Although there are various optical arrangements used for LTRS, most of them are composed of a few key components, as exemplified in one of the earliest LTRS systems shown in Figure 7 [52]. The LTRS system consists of a laser source for trapping and potentially a second laser source for Raman excitation; a microscope to focus the trapping laser beam, image the trapped particle, and collect the Raman signal; a spectrograph/spectrometer or monochromator; and a photoelectronic detector (charge-coupled device (CCD), Image-intensified CCD (ICCD), electron multiplying CCD (EMCCD), photomultiplier tube (PMT), or avalanche photodiode (APD)) to record the Raman spectra. In order to minimize the elastic scattering from the trapping and exciting laser while maximizing the Raman signal, a notch filter, long-pass filter, or dichromatic filter is usually required. Since laser tweezers traps particles near the focal point of the objective lens, the particles are necessarily aligned in the high intensity part of the beam, enabling efficient Raman excitation. As a result, most LTRS systems use the trapping laser to also act as the Raman excitation light source [51,52,57] although a second laser can provide additional flexibility [49,53]. Although photophoretic trapping tends to confine the particle in a low intensity region, Raman spectra have nonetheless been measured with a few second integration time using a single beam to provide both the photophoretic trap and Raman excitation [58].

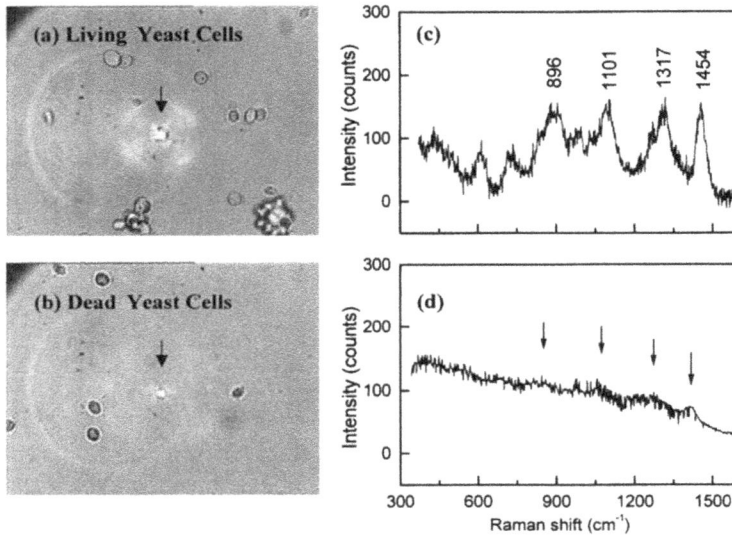

Figure 6. Raman spectra of trapped yeast cells revealing distinct spectra depending on whether the yeast cells are alive or dead [54] (Fair Use according to OSA).

However, some trapping methods have relatively large fluctuations in the particle trapping position (e.g., over a few 10 s of μm), which could reduce the amount of Raman scattered light imaged onto the entrance slit of the spectrometer, making a Raman measurement impractical even with very long integration times (e.g., ~10 s). To overcome such a problem, researchers have introduced a position sensitive detector to monitor the particle position and provide feedback to adjust the laser power in order to hold the particle in a fixed trapping location [58]. Other LTRS systems take advantage of advanced microscopy techniques such as confocal, differential interference contrast, and phase contrast to provide additional functionality or improve the signal to noise ratio of the Raman spectra. For example, combining LTRS with a confocal microscope can efficiently reject out of focus light to improve the Raman signal [3,57,59,60]. Recording phase contrast images in addition to the Raman spectra has provided additional information about the refractility of spores [61,62].

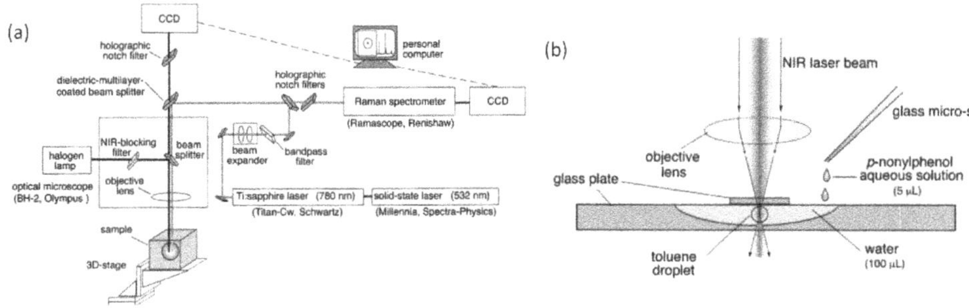

Figure 7. One of the earliest typical LTRS experimental schematics for (**a**) the near-infrared Raman trapping system; and (**b**) the optical arrangement for the sample cell [52] (With permission from ACS publications).

Since multiple particles can be trapped and manipulated simultaneously (e.g., by a holographic or diffractive optical pattern), the combination with LTRS enables longitudinal studies of the interaction between multiple bioaerosols. It also enables studies of the heterogeneous response of different individual particles to a stimuli [57,63,64]. Figure 8a presents a typical schematic of a multifoci-scan confocal Raman imaging system, which relies on a set of galvo-mirrors to rapidly steer the beam in order to generate 40 focal spots under the microscope, each of which is used to trap an individual bacteria spore. The Raman spectra of each of these spores can then be recorded in parallel on an imaging spectrometer by using a third galvo to direct the Raman signal from different particles to different positions along the entrance slit of the spectrometer [57].

3.2. LTRS Studies on Blood Cells

Red blood cells have been frequently studied by LTRS. An early study on the effects of photodamage on trapped red blood cells found that a blood cell trapped with ~2 mW continued to show the same characteristic Raman spectrum for 30 min, whereas a cell trapped with ~20 mW showed a dramatic change in the Raman spectrum after ~15 min, indicating the onset of photodamage [54].

11

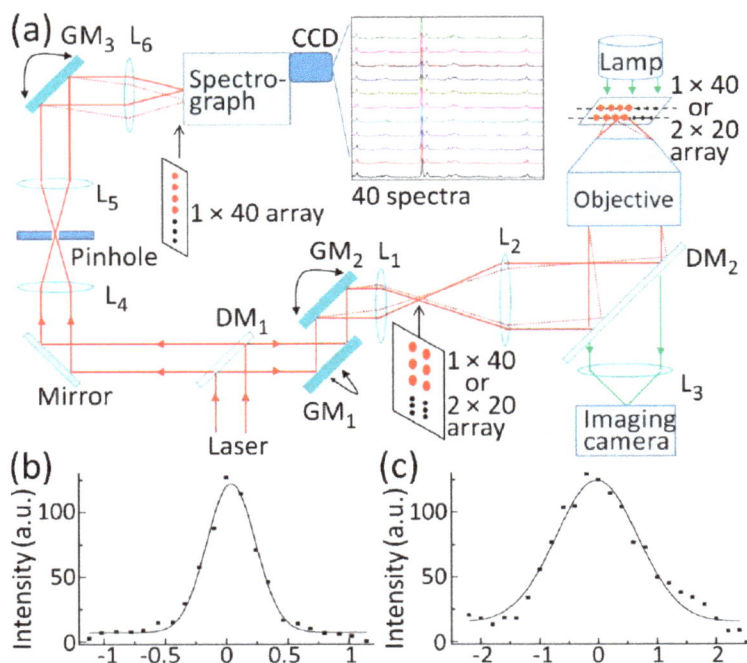

Figure 8. (a) Schematic of a multifoci-scan confocal Raman imaging system; (b) Lateral; and (c) axial intensity profiles of the Raman band at 1001 cm^{-1} of a 100 nm diameter polystyrene bead [57] (With permission from AIP Publishing LLC).

LTRS was used to study the effect of mechanical strain on oxygenation in red blood cells [65]. In this study, the optical tweezers were used to apply a force, stretching a red blood cell, while the Raman spectrum was used to provide a measure of the oxygenation in the cell. The cell was stretched using two optical traps while a third beam was used to provide Raman excitation. Microscope images of the trapped cell before and after mechanical stretching are shown in Figure 9. This study was the first direct measurement of the relationship between optical force and oxygenation, highlighting the unique capabilities of an LTRS system to hold and manipulate a biological particle while simultaneously characterizing its chemical composition.

LTRS has also been used to study blood diseases such as thalassemia [66]. The LTRS study confirmed predictions of reduced oxygenation in thalassemic blood cells and identified differences in the Raman spectra of normal and thalassemic blood cells. In addition, the optical trapping apparatus was used to stretch the blood cells in order to measure the mechanical properties, confirming predictions of increased rigidity in thalassemic blood cells. This highlights the ability of LTRS to study both the chemical and mechanical properties of biological particles in a single experimental setup. LTRS also has potential as a diagnostic tool, and has been shown to differentiate between normal and malaria infected red blood cells [67].

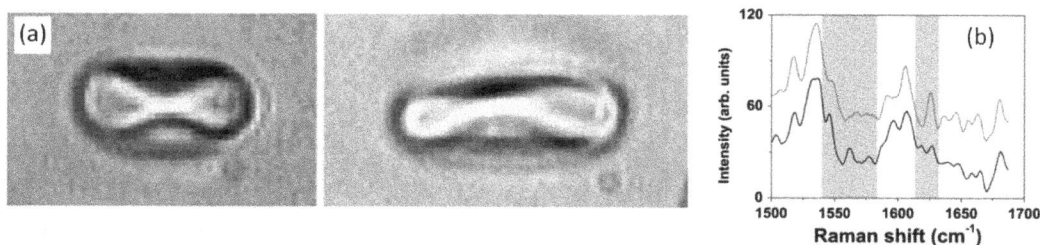

Figure 9. (a) A red blood cell is stretched using optical tweezers while the Raman spectra is monitored to gauge the cell oxygenation level; (b) The Raman spectra of the stretched (bottom curve) and un-stretched (top curve) blood cell. The shaded regions highlight Raman bands which were most affected by mechanical stretching [65] (With permission from Elsevier).

The effects of oxidative stress were studied on red blood cells using LTRS by Zachariah *et al.* [68]. The Raman spectra of 10 normal blood cells and 10 cells exposed to oxidative stress were recorded. Although the Raman spectra exhibited significant cell to cell variation, a PCA of the spectra enabled discrimination between the normal and stressed cells, as shown in Figure 10.

A 2014 study investigated the effect of Ag nanoparticles on red blood cells [69]. Ag nanoparticles have potential anti-microbial applications and are also frequently used in producing surface-enhanced Raman scattering, but their effect in a biological context is not fully understood. Using LTRS, the Raman spectra of trapped red blood cells were measured after exposure to varying concentrations of Ag nanoparticles. As shown in Figure 11, sufficiently high concentrations of Ag nanoparticles altered the relative intensity of the Raman lines at 1211 and 1224 cm^{-1}, corresponding to a change in the methine C-H deformation region of the cell (among other changes) [69]. Monitoring the Raman spectra of exposed blood cells showed that the Ag nanoparticles produce irreversible changes through a transformation from an oxygenated to a de-oxygenated state. By providing information about the temporal evolution of the chemical structure of the blood cells, LTRS provided insights into the cell/nanoparticle interaction process [69].

LTRS has also been performed *in vivo* on red blood cells within the microvessel of a mouse ear [6]. This enabled researchers to measure the relative oxygenation and pH of blood cells in the arterioles and venules. They also compared the Raman spectra of cells measured *in vivo* with cells measured *in vitro* in physiological saline, identifying key differences and highlighting the importance of *in vivo* studies. In addition LTRS enabled a non-destructive measurement without requiring blood extraction [6].

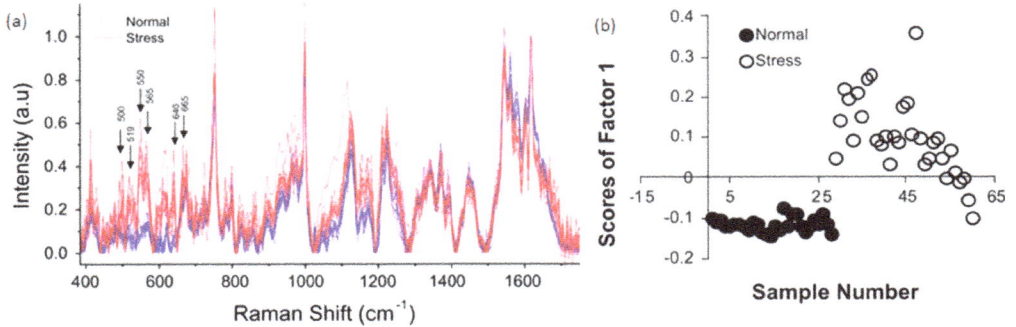

Figure 10. (**a**) Raman spectra from 10 normal cells and 10 cells exposed to oxidative stress. The variations between each spectra illustrate the cell-to-cell variation. Nonetheless, despite broadly similar Raman spectra; a PCA shown in (**b**) clearly differentiates between the stressed and unstressed cells [68] (With permission from Elsevier).

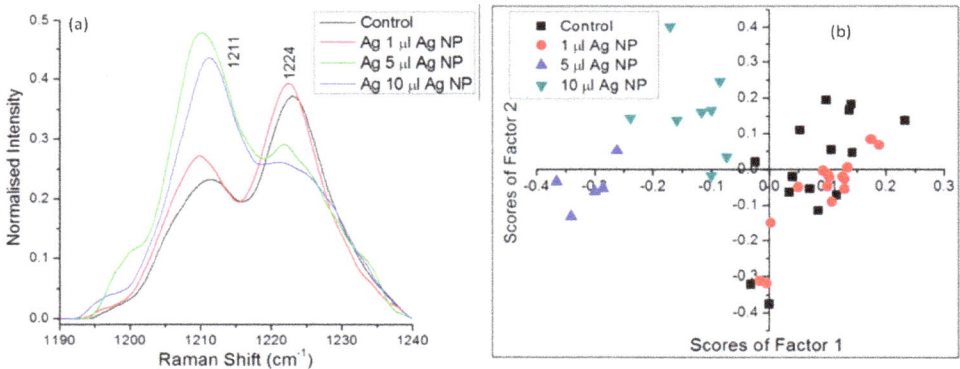

Figure 11. (**a**) LTRS analysis of red blood cells exposed to Ag nanoparticles revealed a change in the relative intensity of the 1211 and 1224 cm^{-1} lines, indicating a change in the methane C-H deformation region of the cell; (**b**) A PCA provided further insight into the temporal evolution of blood cells exposed to Ag nanoparticles [69] (Reprinted under the Creative Commons Attribution License).

3.3. LTRS Studies of Yeast Cells

Yeast cells have also been investigated by several groups using LTRS since a 2002 study revealed differences in the Raman spectra of live and dead trapped yeast cells [54]. Later, a detailed study showed the response of a trapped yeast cell (*Pichia pastoris*) to oxidative stress over time [70]. This result indicated that Raman lines (e.g., 1651 cm^{-1} and 1266 cm^{-1}) associated with C=C stretching and =CH deformation are reduced under exposure to oxidative stress, whereas lines associated with the twisting and bending modes of CH$_2$ remained relatively unaffected. The temporal dependence of varying Raman lines to oxidative stress are shown in Figure 12. The ability of ascorbic acid to mitigate the effects of oxidative stress was also investigated, illustrating the potential of LTRS to evaluate potential therapeutics [70].

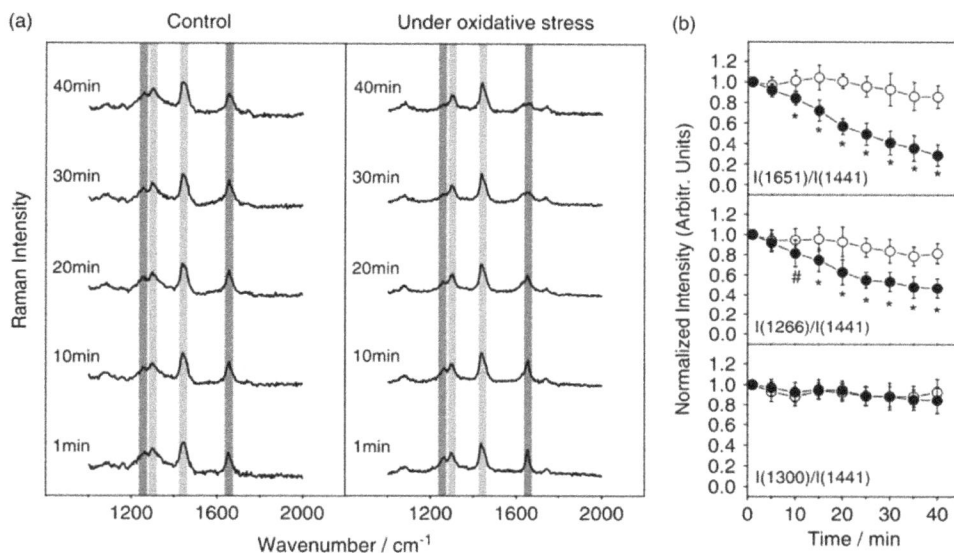

Figure 12. (a) The temporal response of yeast cells to oxidative stress is characterized via LTRS; (b) Raman lines associated with varying chemical bonds within the yeast cell are monitored over time. While the bonds associated with the Raman line at 1651 cm^{-1} and 1441 cm^{-1} are diminished, the bonds associated with the line at 1300 cm^{-1} (among others) are unaffected [70] (With permission of John Wiley & Sons).

3.4. LTRS Studies on Biological and Bacterial Spores

LTRS has been used to study the germination process in *Bacillus* spores by monitoring the time varying calcium dipicolinate biomarker in the Raman spectra [71]. Monitoring numerous individual cells provided information about the variation in the time to germination of individual spores. In addition, the studies of individual cells revealed that the calcium dipicolinate biomarkers were rapidly released in individual spores, albeit at different times for different spores, whereas Raman measurements averaged over a population of spores showed only a smooth decay in the presence of the biomarker [71]. A later study by the same group

15

combined LTRS with measurements of the elastic scattering properties of a *Bacillus* spore during germination to provide additional information about changes in the morphology and refractive index of the spore [61]. As shown in Figure 13, they were able to correlate changes in the elastic scattering of a spore with internal chemical changes monitored via the Raman spectra.

A later study combined LTRS with phase contrast microscopy, providing the first clear demonstration of the correlation between the release of calcium dipicolinate and a change in refractility from bright to dark in the phase contrast images. They found that 70% of the decrease in the intensity of the phase contrast image coincided with the decrease in the calcium dipicolinate Raman line [62]. Additional studies have been performed on the development of *Geobacillus stearothermophilus* spores exposed to varying germinants [72]. LTRS has also been combined with measurements of changes in the speckle pattern formed by light scattered off a trapped cell [59] in a study which compared the dynamics of *E. coli* cells lysed from outside by an egg white lysozyme and from within by a temperature induced bacteriophage. The time varying Raman spectra revealed that the cell underwent significantly different responses in the cases considered. In addition, since the speckle pattern depends sensitively on the morphology of the cell, this provided additional information regarding the release of intracellular materials (e.g., proteins and ribosomes) which disrupted the cell wall.

LTRS has also been used for the identification of bacterial spores in an aqueous environment with a mixture of additional particles [73]. Specifically, the LTRS system was able to identify *Bacillus cereus* spores in a mixed solution of similarly sized polystyrene and silica particles, despite indistinguishable microscope images, as shown in Figure 14. The LTRS-based identification system was validated by sampling 100 particles and found to correctly identify the fraction of each particle type in the mixture. This demonstrated the potential for such an LTRS system as a particle analyzer, possibly in a flow cytometry environment [73]. The LTRS system has the potential for much higher speed particle identification than methods based on cell cultures, and far superior specificity compared with fluorescence based particle identification schemes.

16

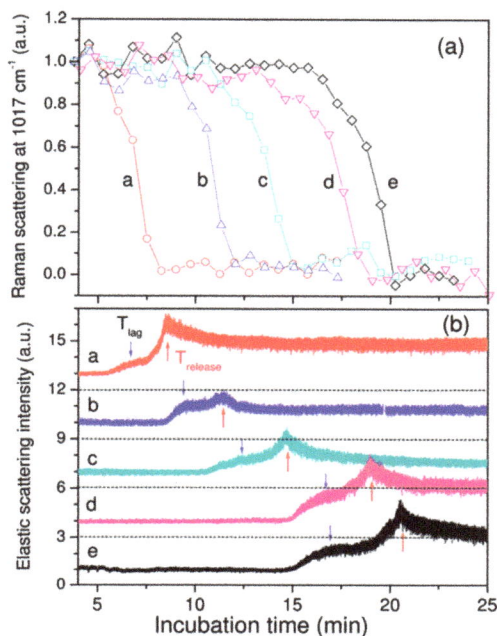

Figure 13. The Raman line (**a**) associated with the calcium dipicolinate biomarker is monitored during the spore germination process along with the intensity of elastic scattered light; (**b**) for varying particles. The individual particles show different germination times, indicated by the rapid decrease in the Raman scattering line at 1017 cm^{-1} (**a**); but the germination process is consistently correlated with an increase in the elastic scattering of the cell (**b**) [61]. (With permission from ACS publications).

3.5. LTRS Used for Drug Discovery and Evaluation

LTRS also has tremendous potential as a tool in the evaluation and understanding of pharmaceuticals. A 2010 study used LTRS to evaluate the response of leukemic T lymphocytes exposed to the chemotherapy drug doxorubicin [8]. Raman spectra were recorded over 72 h after exposure to varying doses of the chemotherapy drug. Raman signatures indicative of changes in vesicle formation, cell membrane blebbing, chromatin condensation, and the cytoplasm of dead cells were observed during varying stages of apoptosis induced by the drug. Due to the heterogeneity in the cellular response, the individual Raman spectrum (shown in Figure 15) is difficult to interpret. However, a PCA was able to clarify the response of cells exposed to varying drug doses. This analysis revealed three distinct stages of apoptosis and the time required for the cell to progress through these stages depended on the drug dose. The ability of LTRS to study individual cells also revealed that certain cells did not respond to the drug and remained in the control

17

group for the duration of the study, indicating that some cells either have a very slow response or exhibit a drug-resistant phenotype. This indicates a potential application of LTRS to rapidly determine if an individual patient will respond to a specific drug treatment. While this initial study recorded the Raman spectrum from a localized position within a cell [8], a follow-up study analyzed the Raman spectrum from the entire cell, further elucidating the cellular response to the drug [74].

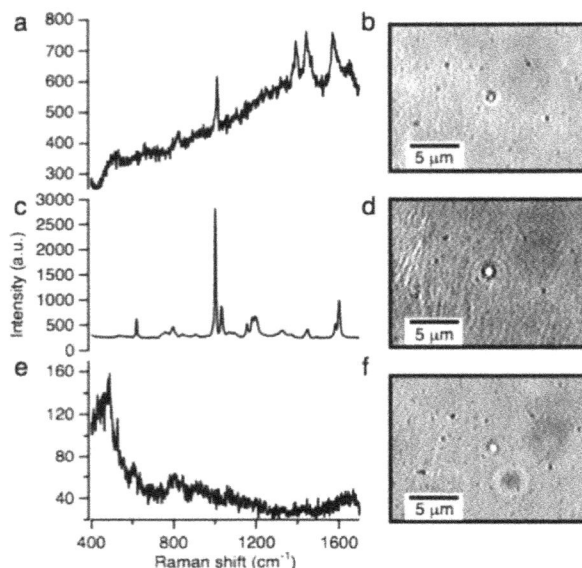

Figure 14. Raman spectra and microscope images of trapped particles of either (**a,b**) *Bacillus cereus* spores; (**c,d**) polystyrene microspheres; (**e,f**) glass microspheres. The LTRS system used the unique Raman spectra to rapidly identify the particle type [73] (With permission from ACS publications).

In addition to characterizing the interaction between cancer cells and potential treatments, LTRS also has potential applications in the identification of cancer cells, as demonstrated in a study on using LTRS to identify epithelial cancer cells [75]. In this study, LTRS was performed on surgically removed human colorectal tissue revealing consistent differences in the Raman spectra of cancerous and non-cancerous cells through PCA.

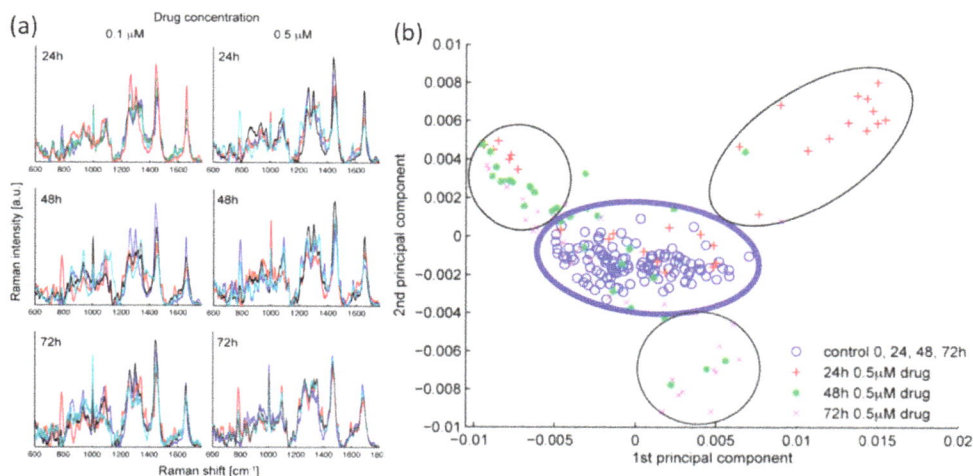

Figure 15. (**a**) Raman spectra recorded from leukemic T lymphocytes exposed to varying doses of a chemotherapy drug over time; (**b**) Principal component analysis revealed three stages in the apoptosis process induced by the drug [8] (Fair Use according to OSA).

3.6. LTRS Studies on Airborne Bioaerosols

There is a high demand for real-time, in-situ detection and characterization of airborne bioaerosols based on single particle Raman spectroscopy. Raman spectroscopy has been shown to be able to discriminate between various microbes and bioaerosols; however, these measurements relied on first collecting the samples on a substrate and the Raman measurements were only performed after collection, limiting the characterization system response time and throughput [76–79]. LTRS has the potential to provide more efficient bioaerosol characterization, without the need for collection on a substrate, which could also interfere with the Raman spectra. However, optical trapping of a micro-particle in air is more challenging than in liquid because of the drag force in air and the larger optical scattering force due to the high refractive index contrast in air [9,24,44]. As a result, there have only been a few studies on airborne bioaerosol particles based on LTRS techniques, although there has been significant progress performing LTRS on airborne droplets which take advantage of the unique optical properties and morphology of airborne droplets. For example, several studies have investigated the effect of morphology-dependent microdroplet resonances on the Raman spectra, as well as studies on phase and size transitions, liquid-gas interactions, thermodynamic behavior, and the kinetics of mass transfer in airborne droplets [40,76–86].

Recently, photophoretic trapping was combined with Raman spectroscopy for the characterization and identification of absorbing bioaerosols, as shown in

19

Figure 16. A 2012 study first presented this technique by measuring the Raman spectra of individual trapped carbon nanotube particles [58]. The Raman spectra of individual airborne carbon nanoclusters have also been measured in a single beam photophoretic trap [87]. A later study reported measurements of Raman spectra from individual bioaerosol particles (pollen particles and fungal spores) held in a photophoretic trap [23]. In these studies the trapping laser also provided the Raman excitation and the distinct Raman spectra could be used for particle discrimination and identification. The photophoretic trap was integrated with an aerosol delivery nozzle to enable efficient particle trapping for potential applications as an on-line aerosol characterization instrument [23,24]. Moreover, such photophoretic traps have been shown to work for a wide range of aerosol types, including biological molecules, proteins, fungal spores, and allergens [34].

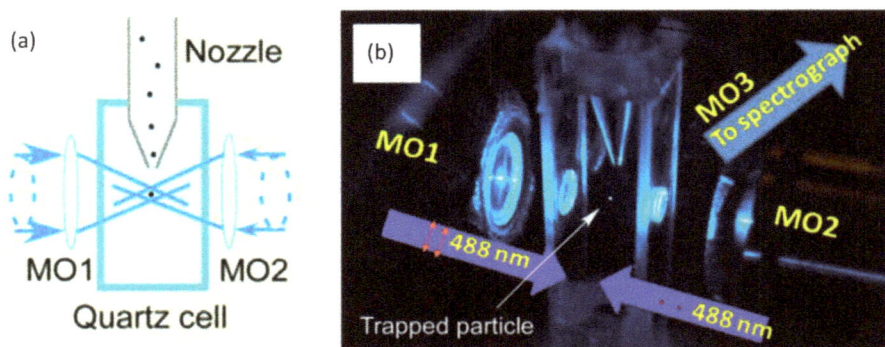

Figure 16. (**a**) LTRS studies on absorbing bioaerosols rely on a photophoretic trap to hold a particle in the low intensity region formed between two counter-propagating hollow beams; (**b**) Photograph of an experimental photophoretic LTRS system in which an aerosol delivery nozzle introduces particles into the photophoretic trap for on-line particle characterization [23]. (With permission from Elsevier).

LTRS has also been applied to clinical and medical research, further illustrating the potential impact of this technique [60,73,74,88,89]. The Raman spectrum is sufficiently sensitive to identify changes in an optically trapped stem cell due to the introduced stress of an attached nanoparticle [89]. In a work by Tong *et al.* [88], particles containing salbutamol, which is the active ingredient in inhalers used by asthmatics, were trapped in air within a high humidity environment to mimic the environment the particle experiences travelling from the inhaler to the lung. Raman spectroscopy was then applied to monitor the molecular changes in the particles as they interacted with the high humidity air while remaining optically trapped. Studies such as this enable a much richer understanding of the delivery of aerosolized pharmaceutical products.

3.7. LTRS in Microfluidics and in Air for Continuously Sampling Bioaerosol Particles

The LTRS techniques we discussed above were primarily applied to study the physical, chemical, or biological properties from one, or a few representative bioaerosol particles, spores, or cells. These single particle studies relied on capturing and trapping individual particles from thousands of potential particles, either by passively waiting for a particle to enter the optical trap, or by actively selecting the particle. However, the ability to continuously trap, characterize, and release individual particles for on-line Raman-based particle identification or for longitudinal studies on a series of successively arriving particles with different properties could further increase the application space for LTRS, particularly if such a system could continuously sample individual particles from a particle stream (e.g., airborne particles from the atmosphere or particles in a liquid reservoir from a patient) over long periods of time [24].

There are two key requirements for such a system: (1) focusing and concentrating the particle stream into a small interrogation volume (e.g., $20 \times 20 \times 20 \ \mu m^3$) through which particles pass sufficiently slowly to be trapped and sampled one at a time; (2) the optical trap needs to be strong enough to capture and hold individual particles from the stream with different optical properties and morphologies. Toward these goals, a method was recently developed to deliver individual particles into the trapping volume based on counter-directional air flow which aerodynamically focused particles into the trapping position with minimal particle loss [24]. A second study showed that a single optical trapping technique could be used for both transparent and absorbing particles regardless of their morphologies [44]. Nevertheless, there has been significantly more success applying LTRS to particles in liquid, particularly in a microfluidic environment. These LTRS systems benefit from the confinement of the microchannel, which efficiently delivers particles to the trapping volume [41,90–96]. In addition to optical trapping, microfluidic systems have also been combined with electrostatic trapping [97–99]. Moreover, combining microfluidics with surface-enhanced Raman scattering (SERS) enables much faster Raman measurements, and it is possible to identify cells and characterize cellular chemical dynamics in flow, without needing to trap the particles [100–105].

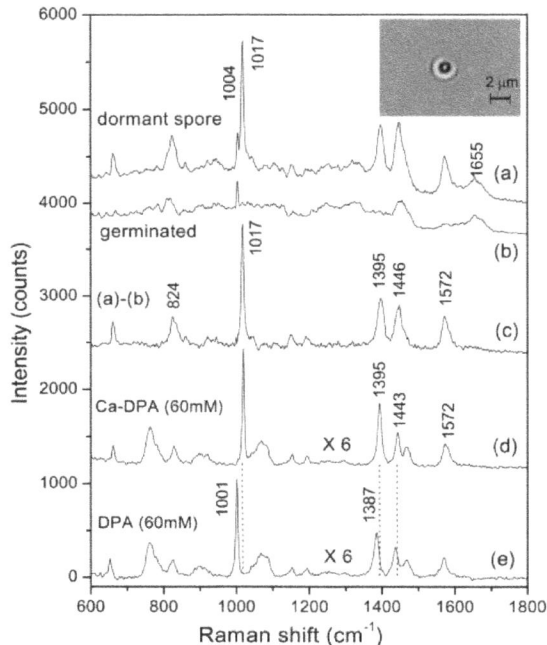

Figure 17. Raman spectra of (**a**) dormant; and (**b**) germinated *B. subtilis* spores; (**c**) subtraction of curve b from curve a; (**d**) The Ca-DPA; and (**e**) DPA Raman spectra [96] (With permission from American Society for Microbiology).

Continuous sampling systems based on LTRS in a microfluidic system tend to resemble the experimental apparatus shown in Figure 4, with the additional ability to store a reservoir of samples in a reservoir or syringe and gradually deliver these particles into the trapping volume. Such a system was shown to discriminate between cancerous and normal cells including erythrocytes, leukocytes, acute myeloid leukemia cells (OCI-AML3), and breast tumor cells BT-20 and MCF-7 [90,91]. In addition, the molecular compositions and structures of single cells or even the subcellular composition could be determined [41]. The physical and chemical mechanisms of many biological processes were also explored based on the interactions between cells as well as between cells and their environment [93]. A label-free cell sorting platform was developed based on intrinsic Raman markers for automated sampling and sorting of a large number of individual cells in solution [94]. Microfluidic based LTRS has been used to monitor the dependence of the neuronal action on nerve globins by obtaining the Raman spectra from several globin-containing cells: hemoglobin (Hb) within single red blood cells, a nerve globin present in the nerve cord of the annelid Aphrodite aculeata (A. aculeata), and wild-type (wt) human neuroglobin (NGB) overexpressed in *Escherichia coli*

(*E. coli*) bacteria [95]. Calcium (Ca) ion-Dipicolinic acid (DPA) levels in individual trapped *Bacillus* spores were measured to provide insight into the spore germination process [96,106]. Ca-DPA is important in spore resistance to environmental stresses and in spore stability, and Ca-DPA levels in spore populations can vary with spore species/strains, as well as with sporulation conditions. Figure 17 shows some representative Raman spectra of single dormant and germinated *Bacillus* spores, as well as the spectra of Ca-DPA, and DPA [96].

4. Conclusions

The combination of optical trapping with Raman spectroscopy has proven itself to be a versatile and powerful tool in the study of biological particles. LTRS enables the measurement of Raman spectra from individual particles for applications ranging from particle detection and identification to longitudinal studies of the response of a biological particle such as a cell to environmental changes. Optical tweezers enable the localized measurement of Raman spectra from varying positions within a cell, as well as providing multi-modality cell characterization by using, for example, the optical tweezers to measure the mechanical properties of a particle while the Raman spectrum provides information about the chemical makeup of the particle. The ability to study individual particles, as opposed to collecting the combined Raman spectrum from a population of cells, provides additional information about the heterogeneity of the cells and the variation in the cell-to-cell response to environmental changes. This unique functionality has enabled researchers to identify the fast release of calcium dipicolinate in yeast cells, as well as to identify the fraction of cancer cells which respond to chemotherapy.

Although optical trapping holds particles in place long enough to make Raman measurements possible, the long exposure time still imposes a limitation on the throughput of LTRS systems. This limited throughput could be particularly challenging in on-line particle characterization techniques which use LTRS to identify airborne biological particles or cells in a microfluidic environment. As a result, a promising new area of research combines advances in stimulated Raman scattering or coherent anti-stokes Raman measurements with optical trapping.

Acknowledgments: We acknowledge support by the Defense Threat Reduction Agency (DTRA) under contract number HDTRA 1514122 and HDTRA1310184, US Army Research Laboratory mission funds, and under Cooperative Agreement Number W911NF-12-2-0019.

Conflicts of Interest: The authors declare no conflict of interest.

References

1. Camp, C.H., Jr.; Cicerone, M.T. Chemically sensitive bioimaging with coherent Raman scattering. *Nat. Photonics* **2015**, *9*, 295–305.

2. Chan, J.W. Recent advances in laser tweezers Raman spectroscopy (LTRS) for label-free analysis of single cells. *J. Biophoton.* **2013**, *6*, 36–48.

3. Puppels, G.J.; de Mul, F.F.; Otto, C.; Greve, J.; Robert-Nicoud, M.; Arndt-Jovin, D.J.; Jovin, T.M. Studying single living cells and chromosomes by confocal Raman microspectroscopy. *Nature* **1990**, *347*, 301–303.

4. Thurn, R.; Kiefer, W. Raman-Microsampling Technique Applying Optical Levitation by Radiation Pressure. *Appl. Spectrosc.* **1984**, *38*, 78–83.

5. Xie, C.; Li, Y.Q.; Tang, W.; Newton, R.J. Study of dynamical process of heat denaturation in optically trapped single microorganisms by near-infrared Raman spectroscopy. *J. Appl. Phys.* **2003**, *94*, 6138–6142.

6. Shao, J.; Yao, H.; Meng, L.; Li, Y.; Lin, M.; Li, X.; Liu, J.; Liang, J. Raman spectroscopy of circulating single red blood cells in microvessels *in vivo*. *Vib. Spectrosc.* **2012**, *63*, 367–370.

7. Fu, D.; Zhou, J.; Zhu, W.S.; Manley, P.W.; Wang, Y.K.; Hood, T.; Wylie, A.; Xie, X.S. Imaging the intracellular distribution of tyrosine kinase inhibitors in living cells with quantitative hyperspectral stimulated Raman scattering. *Nat. Chem.* **2014**, *6*, 614–622.

8. Moritz, T.J.; Taylor, D.S.; Krol, D.M.; Fritch, J.; Chan, J.W. Detection of doxorubicin-induced apoptosis of leukemic T-lymphocytes by laser tweezers Raman spectroscopy. *Biomed. Opt. Express* **2010**, *1*, 1138–1147.

9. Li, T. *Fundamental Tests of Physics with Optically Trapped Microspheres*; Springer Theses: New York, NY, USA, 2013.

10. Wills, J.B.; Knox, K.J.; Reid, J.P. Optical control and characterisation of aerosol. *Chem. Phys. Lett.* **2009**, *481*, 153–165.

11. Ashkin, A. Acceleration and trapping of particles by radiation pressure. *Phys. Rev. Lett.* **1970**, *24*, 156–159.

12. Lewittes, M.; Arnold, S.; Oster, G. Radiometric levitation of micron sized spheres Radiometric levitation of micron sized spheres. *Appl. Phys. Lett.* **1982**, *40*, 455–457.

13. Ashkin, A.; Dziedzic, J.M. Optical levitation by radiation pressure. *Appl. Phys. Lett.* **1971**, *19*, 283–285.

14. Roosen, G.; Imbert, C. Optical levitation by means of two horizontal laser beams: A theoretical and experimental study. *Phys. Lett. A* **1976**, *59*, 6–8.

15. Omori, R.; Kobayashi, T.; Suzuki, A. Observation of a single-beam gradient-force optical trap for dielectric particles in air. *Opt. Lett.* **1997**, *22*, 816–818.

16. Neuman, K.C.; Block, S.M. Optical trapping. *Rev. Sci. Instrum.* **2004**, *75*, 2787–2809.

17. Burnham, D.R.; McGloin, D. Modeling of optical traps for aerosols. *J. Opt. Soc. Am. B* **2011**, *28*, 2856–2864.

18. Agate, B.; Brown, C.; Sibbett, W.; Dholakia, K. Femtosecond optical tweezers for in-situ control of two-photon fluorescence. *Opt. Express* **2004**, *12*, 3011–3017.

19. Gahagan, K.T.; Swartzlander, G. A Optical vortex trapping of particles. *Opt. Lett.* **1996**, *21*, 827–829.

20. Desyatnikov, A.S.; Shvedov, V.G.; Rode, A.V.; Krolikowski, W.; Kivshar, Y.S. Photophoretic manipulation of absorbing aerosol particles with vortex beams: Theory versus experiment. *Opt. Express* **2009**, *17*, 8201–8211.

21. Shvedov, V.G.; Desyatnikov, A.S.; Rode, A.V.; Krolikowski, W.; Kivshar, Y.S. Optical guiding of absorbing nanoclusters in air. *Opt. Express* **2009**, *17*, 5743–5757.

22. Shvedov, V.; Davoyan, A.R.; Hnatovsky, C.; Engheta, N.; Krolikowski, W. A long-range polarization-controlled optical tractor beam. *Nat. Photonics* **2014**, *8*, 846–850.

23. Wang, C.; Pan, Y.; Hill, S.C.; Redding, B. Photophoretic trapping-Raman spectroscopy for single pollens and fungal spores trapped in air. *J. Quant. Spectrosc. Radiat. Transf.* **2015**, *153*, 4–12.

24. Pan, Y.L.; Wang, C.; Hill, S.C.; Coleman, M.; Beresnev, L.A.; Santarpia, J.L. Trapping of individual airborne absorbing particles using a counterflow nozzle and photophoretic trap for continuous sampling and analysis. *Appl. Phys. Lett.* **2014**, *104*, 113507.

25. Liu, F.; Zhang, Z.; Wei, Y.; Zhang, Q.; Cheng, T.; Wu, X. Photophoretic trapping of multiple particles in tapered-ring optical field. *Opt. Express* **2014**, *22*, 23716–23723.

26. Shvedov, V.G.; Hnatovsky, C.; Shostka, N.; Rode, A.V.; Krolikowski, W. Optical manipulation of particle ensembles in air. *Opt. Lett.* **2012**, *37*, 1934–1936.

27. Zhang, P.; Zhang, Z.; Prakash, J.; Huang, S.; Hernandez, D.; Salazar, M.; Christodoulides, D.N.; Chen, Z. Trapping and transporting aerosols with a single optical bottle beam generated by moiré techniques. *Opt. Lett.* **2011**, *36*, 1491–1493.

28. Shvedov, V.G.; Rode, A.V; Izdebskaya, Y.V.; Desyatnikov, A.S.; Krolikowski, W.; Kivshar, Y.S. Selective trapping of multiple particles by volume speckle field. *Opt. Express* **2010**, *18*, 3137–3142.

29. Zhang, Z.; Cannan, D.; Liu, J.; Zhang, P.; Christodoulides, D.N.; Chen, Z. Observation of trapping and transporting air-borne absorbing particles with a single optical beam. *Opt. Express* **2012**, *20*, 16212–16217.

30. Lin, J.; Li, Y. Optical trapping and rotation of airborne absorbing particles with a single focused laser beam. *Appl. Phys. Lett.* **2014**, *104*, 101909.

31. Rohatschek, H. Direction Magnitude and Causes of Photoporetic Forces. *J. Aerosol Sci.* **1985**, *16*, 29–42.

32. Rohatschek, H. Semi-empirical model of photophoretic forces for the entire range of pressures. *J. Aerosol Sci.* **1995**, *26*, 717–734.

33. Jovanovic, O. Photophoresis—Light induced motion of particles suspended in gas. *J. Quant. Spectrosc. Radiat. Transf.* **2009**, *110*, 889–901.

34. Redding, B.; Hill, S.C.; Alexson, D.; Wang, C.; Pan, Y. Photophoretic trapping of airborne particles using ultraviolet illumination. *Opt. Express* **2015**, *23*, 3630–3639.

35. Lin, J.; Hart, A.G.; Li, Y. Optical pulling of airborne absorbing particles and smut spores over a meter-scale distance with negative photophoretic force. *Appl. Phys. Lett.* **2015**, *106*, 171906.

36. Dufresne, E.R.; Spalding, G.C.; Dearing, M.T.; Sheets, S.A.; Grier, D.G. Computer-generated holographic optical tweezer arrays. *Rev. Sci. Instrum.* **2001**, *72*, 1810–1816.

37. Dufresne, E.R.; Grier, D.G. Optical tweezer arrays and optical substrates created with diffractive optics. *Rev. Sci. Instrum.* **1998**, *69*, 1974–1977.

38. Melville, H.; Milne, G.; Spalding, G.; Sibbett, W.; Dholakia, K.; McGloin, D. Optical trapping of three-dimensional structures using dynamic holograms. *Opt. Express* **2003**, *11*, 3562–3567.

39. Emiliani, V.; Cojoc, D.; Ferrari, E.; Garbin, V.; Durieux, C.; Coppey-Moisan, M.; di Fabrizio, E. Wave front engineering for microscopy of living cells. *Opt. Express* **2005**, *13*, 1395–1405.

40. Burnham, D.R.; McGloin, D. Holographic optical trapping of aerosol droplets. *Opt. Express* **2006**, *14*, 4175–4181.

41. Jess, P.R.T.; Garcés-Chávez, V.; Smith, D.; Mazilu, M.; Paterson, L.; Riches, A.; Herrington, C.S.; Sibbett, W.; Dholakia, K. Dual beam fibre trap for Raman micro-spectroscopy of single cells. *Opt. Express* **2006**, *14*, 5779–5791.

42. Čižmár, T.; Dholakia, K. Shaping the light transmission through a multimode optical fibre: Complex transformation analysis and applications in biophotonics. *Opt. Express* **2011**, *19*, 18871–18884.

43. Bianchi, S.; Di Leonardo, R. A multi-mode fiber probe for holographic micromanipulation and microscopy. *Lab Chip* **2012**, *12*, 635–639.

44. Redding, B.; Pan, Y.L. Optical trap for both transparent and absorbing particles in air using a single shaped laser beam. *Opt. Lett.* **2015**, *40*, 2798–2801.

45. Dear, R.D.; Burnham, D.R.; Summers, M.D.; McGloin, D.; Ritchie, G.A.D. Single aerosol trapping with an annular beam: Improved particle localisation. *Phys. Chem. Chem. Phys.* **2012**, *14*, 15826–15831.

46. Beams, J.W. Production and use of high centrifugal fields. *Science* **1954**, *120*, 619–625.

47. Wuerker, R.F.; Shelton, H.; Langmuir, R.V. Electrodynamic containment of charged particles. *J. Appl. Phys.* **1959**, *30*, 342–349.

48. Wu, J.; Du, G. Acoustic radiation force on a small compressible sphere in focused beam. *J. Acoust. Soc. Am.* **1990**, *87*, 997–1003.

49. Knoll, P.; Marchl, M.; Kiefer, W. Raman Spectroscopy of Microparticles in Laser Light Traps. *Indian J. Pure Appl. Phys.* **1988**, *26*, 268–277.

50. Lankers, M.; Popp, J.; Kiefer, W. Raman and fluorescence spectra of single optically trapped microdroplets in emulsions. *Appl. Spectrosc.* **1994**, *48*, 1166–1168.

51. Ajito, K. Combined Near-Infrared Raman Microprobe and Laser Trapping System: Application to the Analysis of a Single Organic Microdroplet in Water. *Appl. Spectrosc.* **1998**, *52*, 339–342.

52. Ajito, K.; Morita, M.; Torimitsu, K. Investigation of the molecular extraction process in single subpicoliter droplets using a near-infrared laser Raman trapping system. *Anal. Chem.* **2000**, *72*, 4721–4725.

53. Ajito, K.; Torimitsu, K. Laser trapping and Raman spectroscopy of single cellular organelles in the nanometer range. *Lab Chip* **2002**, *2*, 11–14.

54. Xie, C.; Dinno, M.A.; Li, Y.Q. Near-infrared Raman spectroscopy of single optically trapped biological cells. *Opt. Lett.* **2002**, *27*, 249–251.

55. Alexander, T.A.; Pellegrino, P.; Gillespie, J.B. Near-infrared Surface-Enhanced-Raman-Scattering (SERS) mediated detection of single, optically trapped, bacterial spores. *Appl. Spectrosc.* **2003**, *57*, 1340–1345.

56. Xie, C.; Goodman, C.; Dinno, M.; Li, Y.Q. Real-time Raman spectroscopy of optically trapped living cells and organelles. *Opt. Express* **2004**, *12*, 6208–6214.

57. Kong, L.; Zhang, P.; Yu, J.; Setlow, P.; Li, Y.Q. Rapid confocal Raman imaging using a synchro multifoci-scan scheme for dynamic monitoring of single living cells. *Appl. Phys. Lett.* **2011**, *98*, 4–7.

58. Pan, Y.L.; Hill, S.C.; Coleman, M. Photophoretic trapping of absorbing particles in air and measurement of their single-particle Raman spectra. *Opt. Express* **2012**, *20*, 5325–5334.

59. Chen, D.; Shelenkova, L.; Li, Y.; Kempf, C.R.; Sabelnikov, A. Laser tweezers Raman spectroscopy potential for studies of complex dynamic cellular processes: Single cell bacterial analysis. *Anal. Chem.* **2009**, *81*, 3227–3238.

60. Houlne, M.P.; Sjostrom, C.M.; Uibel, R.H.; Kleimeyer, J.A.; Harris, J.M. Confocal Raman microscopy for monitoring chemical reactions on single optically trapped, solid-phase support particles. *Anal. Chem.* **2002**, *74*, 4311–4319.

61. Peng, L.; Chen, D.; Setlow, P.; Li, Y.Q. Elastic and inelastic light scattering from single bacterial spores in an optical trap allows the monitoring of spore germination dynamics. *Anal. Chem.* **2009**, *81*, 4035–4042.

62. Kong, L.; Zhang, P.; Setlow, P.; Li, Y.Q. Characterization of bacterial spore germination using integrated phase contrast microscopy, Raman spectroscopy, and optical tweezers. *Anal. Chem.* **2010**, *82*, 3840–3847.

63. Trivedi, R.P.; Lee, T.; Bertness, K.A.; Smalyukh, I.I. Three dimensional optical manipulation and structural imaging of soft materials by use of laser tweezers and multimodal nonlinear microscopy. *Opt. Express* **2010**, *18*, 27658–27669.

64. Butler, J.R.; Wills, J.B.; Mitchem, L.; Burnham, D.R.; McGloin, D.; Reid, J.P. Spectroscopic characterisation and manipulation of arrays of sub-picolitre aerosol droplets. *Lab Chip* **2009**, *9*, 521–528.

65. Rao, S.; Bálint, S.; Cossins, B.; Guallar, V.; Petrov, D. Raman study of mechanically induced oxygenation state transition of red blood cells using optical tweezers. *Biophys. J.* **2009**, *96*, 209–216.

66. De Luca, A.C.; Rusciano, G.; Ciancia, R.; Martinelli, V.; Pesce, G.; Rotoli, B.; Selvaggi, L.; Sasso, A. Spectroscopical and mechanical characterization of normal and thalassemic red blood cells by Raman Tweezers. *Opt. Express* **2008**, *16*, 7943–7957.

67. Dasgupta, R.; Verma, R.S.; Ahlawat, S.; Uppal, A.; Gupta, P.K. Studies on erythrocytes in malaria infected blood sample with Raman optical tweezers. *J. Biomed. Opt.* **2011**, *16*, 077009.

68. Zachariah, E.; Bankapur, A.; Santhosh, C.; Valiathan, M.; Mathur, D. Probing oxidative stress in single erythrocytes with Raman Tweezers. *J. Photochem. Photobiol. B Biol.* **2010**, *100*, 113–116.

69. Bankapur, A.; Barkur, S.; Chidangil, S.; Mathur, D. A micro-raman study of live, single Red Blood Cells (RBCs) treated with AgNO3 nanoparticles. *PLoS ONE* **2014**, *97*, e103493.

70. Chang, W.T.; Lin, H.L.; Chen, H.C.; Wu, Y.M.; Chen, W.J.; Lee, Y.T.; Liau, I. Real-time molecular assessment on oxidative injury of single cells using Raman spectroscopy. *J. Raman Spectrosc.* **2009**, *40*, 1194–1199.

71. Chen, D.; Huang, S.S.; Li, Y.Q. Real-time detection of kinetic germination and heterogeneity of single Bacillus spores by laser tweezers Raman spectroscopy. *Anal. Chem.* **2006**, *78*, 6936–6941.

72. Zhou, T.; Dong, Z.; Setlow, P.; Li, Y.Q. Kinetics of Germination of Individual Spores of Geobacillus stearothermophilus as Measured by Raman Spectroscopy and Differential Interference Contrast Microscopy. *PLoS ONE* **2013**, *8*, 1–11.

73. Chan, J.W.; Esposito, A.P.; Talley, C.E.; Hollars, C.W.; Lane, S.M.; Huser, T. Reagentless Identification of Single Bacterial Spores in Aqueous Solution by Confocal Laser Tweezers Raman Spectroscopy. *Anal. Chem.* **2004**, *76*, 599–603.

74. Schie, I.W.; Alber, L.; Gryshuk, A.L.; Chan, J.W. Investigating drug induced changes in single, living lymphocytes based on Raman micro-spectroscopy. *Analyst* **2014**, *2726*, 2726–2733.

75. Chen, K.; Qin, Y.; Zheng, F.; Sun, M.; Shi, D. Diagnosis of colorectal cancer using Raman spectroscopy of laser-trapped single living epithelial cells. *Opt. Lett.* **2006**, *31*, 2015–2017.

76. Rosch, P.; Harz, M.; Peschke, K.; Ronneberger, O.; Burkhardt, H. On-Line Monitoring and Identification of. *Anal. Chem.* **2006**, *78*, 2163–2170.

77. Schulte, F.; Lingott, J.; Panne, U.; Kneipp, J. Chemical Characterization and Classification of Pollen Chemical Characterization and Classification of Pollen. *Anal. Chem.* **2008**, *80*, 9551–9556.

78. Kalasinsky, K.S.; Hadfield, T.; Shea, A.A.; Kalasinsky, V.F.; Nelson, M.P.; Neiss, J.; Drauch, A.J.; Vanni, G.S.; Treado, P.J. Raman chemical imaging spectroscopy reagentless detection and identification of pathogens: Signature development and evaluation. *Anal. Chem.* **2007**, *79*, 2658–2673.

79. De Gelder, J.; de Gussem, K.; Vandenabeele, P.; Moens, L. Reference database of Raman spectra of biological molecules. *J. Raman Spectrosc.* **2007**, *38*, 1133–1147.

80. Schaschek, K.; Popp, J.; Kiefer, W. Observation of morphology-dependent input and output resonances in time-dependent Raman spectra of optically levitated microdroplets. *J. Raman Spectrosc.* **1993**, *24*, 69–75.

81. Kaiser, T.; Roll, G.; Schweiger, G. Investigation of coated droplets in an optical trap: Raman-scattering, elastic-light-scattering, and evaporation characteristics. *Appl. Opt.* **1996**, *35*, 5918–5924.

82. King, M.D.; Thompson, K.C.; Ward, A.D. Laser tweezers raman study of optically trapped aerosol droplets of seawater and oleic acid reacting with ozone: Implications for cloud-droplet properties. *J. Am. Chem. Soc.* **2004**, *126*, 16710–16711.

83. Symes, R.; Sayer, R.M.; Reid, J.P. Cavity enhanced droplet spectroscopy: Principles, perspectives and prospects. *Phys. Chem. Phys. Chem.* **2004**, *6*, 474–487.

84. Mitchem, L.; Buajarern, J.; Hopkins, R.J.; Ward, A.D.; Gilham, R.J.J.; Johnston, R.L.; Reid, J.P. Spectroscopy of growing and evaporating water droplets: Exploring the variation in equilibrium droplet size with relative humidity. *J. Phys. Chem. A* **2006**, *110*, 8116–8125.

85. Butler, J.R.; Mitchem, L.; Hanford, K.L.; Treuel, L.; Reid, J.P. In situ comparative measurements of the properties of aerosol droplets of different chemical composition. *Faraday Discuss.* **2008**, *137*, 351–366.

86. Kriek, M.; Neylon, C.; Roach, P.L.; Clark, I.P.; Parker, A.W. A simple setup for the study of microvolume frozen samples using Raman spectroscopy. *Rev. Sci. Instrum.* **2005**, *76*, 1–3.

87. Ling, L.; Li, Y. Measurement of Raman spectra of single airborne absorbing particles trapped by a single laser beam. *Opt. Lett.* **2013**, *38*, 416–418.

88. Tong, H.J.; Fitzgerald, C.; Gallimore, P.J.; Kalberer, M.; Kuimova, M.K.; Seville, P.C.; Ward, A.D.; Pope, F.D. Rapid interrogation of the physical and chemical characteristics of salbutamol sulphate aerosol from a pressurised metered-dose inhaler (pMDI). *Chem. Commun.* **2014**, *50*, 15499–15502.

89. Bankapur, A.; Krishnamurthy, R.S.; Zachariah, E.; Santhosh, C.; Chougule, B.; Praveen, B.; Valiathan, M.; Mathur, D. Micro-raman spectroscopy of silver nanoparticle induced stress on optically-trapped stem cells. *PLoS ONE* **2012**, *7*, e35075.

90. Pallaoro, A.; Hoonejani, M.R.; Braun, G.B.; Meinhart, C.; Moskovits, M. Combined surface-enhanced Raman spectroscopy biotags and microfluidic platform for quantitative ratiometric discrimination between noncancerous and cancerous cells in flow. *J. Nanophotonics* **2013**, *7*, 073092.

91. Dochow, S.; Krafft, C.; Neugebauer, U.; Bocklitz, T.; Henkel, T.; Mayer, G.; Albert, J.; Popp, J. Tumour cell identification by means of Raman spectroscopy in combination with optical traps and microfluidic environments. *Lab Chip* **2011**, *11*, 1484–1490.

92. Jess, P.R.T.; Garcés-Chávez, V.; Riches, A.C.; Herrington, C.S.; Dholakia, K. Simultaneous Raman micro–spectroscopy of optically trapped and stacked cells. *J. Raman Spectrosc.* **2007**, *38*, 1082–1088.

93. Ramser, K.; Enger, J.; Goksör, M.; Hanstorp, D.; Logg, K.; Käll, M. A microfluidic system enabling Raman measurements of the oxygenation cycle in single optically trapped red blood cells. *Lab Chip* **2005**, *5*, 431–436.

94. Lau, A.Y.; Lee, L.P.; Chan, J.W. An integrated optofluidic platform for Raman-activated cell sorting. *Lab Chip* **2008**, *8*, 1116–1120.

95. Ramser, K.; Wenseleers, W.; Dewilde, S.; van Doorslaer, S.; Moens, L. The combination of resonance Raman spectroscopy, optical tweezers and microfluidic systems applied to the study of various heme-containing single cells. *Spectroscopy* **2008**, *22*, 287–295.

96. Huang, S.S.; Chen, D.; Pelczar, P.L.; Vepachedu, V.R.; Setlow, P.; Li, Y.Q. Levels of Ca^{2+}-dipicolinic acid in individual Bacillus spores determined using microfluidic Raman tweezers. *J. Bacteriol.* **2007**, *189*, 4681–4687.

97. Lesser-Rojas, L.; Ebbinghaus, P.; Vasan, G.; Chu, M.L.; Erbe, A.; Chou, C.F. Low-copy number protein detection by electrode nanogap-enabled dielectrophoretic trapping for surface-enhanced Raman spectroscopy and electronic measurements. *Nano Lett.* **2014**, *14*, 2242–2250.

98. Deng, Y.-L.; Juang, Y.-J. Electrokinetic trapping and surface enhanced Raman scattering detection of biomolecules using optofluidic device integrated with a microneedles array. *Biomicrofluidics* **2013**, *7*, 014111.

99. Perozziello, G.; Catalano, R.; Francardi, M.; Rondanina, E.; Pardeo, F.; De Angelis, F.; Malara, N.; Candeloro, P.; Morrone, G.; di Fabrizio, E. A microfluidic device integrating plasmonic nanodevices for Raman spectroscopy analysis on trapped single living cells. *Microelectron. Eng.* **2013**, *111*, 314–319.

100. Pallaoro, A.; Hoonejani, M.R.; Braun, G.B.; Meinhart, C.D.; Moskovits, M. Rapid Identification by Surface-Enhanced Raman Spectroscopy of Cancer Cells at Low Concentrations Flowing in a Microfluidic Channel. *ACS Nano* **2015**, *4*, 4328–4336.

101. Li, M.; Zhao, F.; Zeng, J.; Qi, J.; Lu, J.; Shih, W.C. Microfluidic surface-enhanced Raman scattering sensor with monolithically integrated nanoporous gold disk arrays for rapid and label-free biomolecular detection. *J. Biomed. Opt.* **2014**, *19*, 111611.

102. Kho, K.W.; Qing, K.Z.M.; Shen, Z.X.; Ahmad, I.B.; Lim, S.S.C.; Mhaisalkar, S.; White, T.J.; Watt, F.; Soo, K.C.; Olivo, M. Polymer-based microfluidics with surface-enhanced Raman-spectroscopy-active periodic metal nanostructures for biofluid analysis. *J. Biomed. Opt.* **2015**, *13*, 054026.

103. Zhang, X.; Yin, H.; Cooper, J.M.; Haswell, S.J. Characterization of cellular chemical dynamics using combined microfluidic and Raman techniques. *Anal. Bioanal. Chem.* **2008**, *390*, 833–840.

104. Wilson, R.; Monaghan, P.; Bowden, S.A.; Parnell, J.; Cooper, J.M. Surface-enhanced raman signatures of pigmentation of cyanobacteria from within geological samples in a spectroscopic-microfluidic flow cell. *Anal. Chem.* **2007**, *79*, 7036–7041.

105. Liu, G.L.; Lee, L.P. Nanowell surface enhanced Raman scattering arrays fabricated by soft-lithography for label-free biomolecular detections in integrated microfluidics. *Appl. Phys. Lett.* **2005**, *87*, 074101.

106. Wang, S.; Setlow, P.; Li, Y. Slow Leakage of Ca-Dipicolinic Acid from Individual Bacillus Spores during Initiation of Spore Germination. *J. Bacteriol.* **2015**, *197*, 1095–1103.

Label-Free Biosensor Imaging on Photonic Crystal Surfaces

Yue Zhuo and Brian T. Cunningham

Abstract: We review the development and application of nanostructured photonic crystal surfaces and a hyperspectral reflectance imaging detection instrument which, when used together, represent a new form of optical microscopy that enables label-free, quantitative, and kinetic monitoring of biomaterial interaction with substrate surfaces. Photonic Crystal Enhanced Microscopy (PCEM) has been used to detect broad classes of materials which include dielectric nanoparticles, metal plasmonic nanoparticles, biomolecular layers, and live cells. Because PCEM does not require cytotoxic stains or photobleachable fluorescent dyes, it is especially useful for monitoring the long-term interactions of cells with extracellular matrix surfaces. PCEM is only sensitive to the attachment of cell components within ~200 nm of the photonic crystal surface, which may correspond to the region of most interest for adhesion processes that involve stem cell differentiation, chemotaxis, and metastasis. PCEM has also demonstrated sufficient sensitivity for sensing nanoparticle contrast agents that are roughly the same size as protein molecules, which may enable applications in "digital" diagnostics with single molecule sensing resolution. We will review PCEM's development history, operating principles, nanostructure design, and imaging modalities that enable tracking of optical scatterers, emitters, absorbers, and centers of dielectric permittivity.

Reprinted from *Sensors*. Cite as: Zhuo, Y.; Cunningham, B.T. Label-Free Biosensor Imaging on Photonic Crystal Surfaces. *Sensors* **2015**, *15*, 21613–21635.

1. Introduction

A photonic crystal (PC) surface is a periodic-modulated dielectric nano-structure material (one example can be seen in Figure 1A) [1–5]. PC surfaces can be designed to provide photonic bandgaps (Figure 1B), within which light propagation is prohibited for specific wavelengths [6–8]. Therefore, the local optical modes provided by the PC surface can be utilized as a highly sensitive and label-free platform for biosensing and bioimaging in life science research. PC surface biosensors [9–30] have been widely used to detect refractive index changes induced by surface-attached biomaterials (Figure 1C,D), and for analytes spanning a wide range of dimensions, including small molecules [31–35], virus particles [36], DNA microarrays [37], and live cells [38–45]. Generally, biosensing is realized with a transducer surface (e.g., PC surface, waveguide or microcavity) and an instrument for collecting the average response from the entire sensing area. When spatially resolved information is required, such as the behavior within individual cells, it is necessary to measure localized responses that can be differentiated from neighboring locations. Thus, spatial resolution becomes a critical factor for biosensor imaging. Among the earliest developed label-free imaging modalities based on PC biosensors [12,16,38,46,47], Photonic Crystal Enhanced Microscopy (PCEM) [12,38,44–51] represents a new form of optical microscopy that uses a PC surface to dynamically detect and visualize biomaterial-surface interactions (Figures 2–4). Because the detection is label-free, it is not limited by the transient activity of fluorescent contrast agents that may be limited by photobleaching effects. Hence, PCEM can be performed for extended time periods to enable study of cell functions (including cell adhesion, migration, apoptosis, and differentiation) that take place over the course of several hours or multiple days.

Based on the number of directions with a periodic repetition of refractive index (RI) contrast, PC nano-structures can be categorized as one-dimensional (1D), two-dimensional (2D), or three-dimensional (3D). A PC surface typically consists of an area of continuous 1D or 2D PC structure on the substrate surface. Here we describe the case of a 1D PC structure as an example to explain label-free biosensor imaging on PC surfaces. Traditionally, a 1D PC is characterized as a multilayer stack of materials with alternating dielectric constants, which are also referred to as Bragg mirrors (or dielectric mirrors) [52–59]. In such a 1D PC stack, the periodicity is normal to the substrate plane and a photonic bandgap is formed for light with the evanescent part of the wavevector (which is highly sensitive to surface RI modifications) normal to the substrate surface. When used in biosensing and bioimaging, this PC structure utilizes the surface electromagnetic waves bound to the multilayer (named Bloch surface waves or surface electromagnetic waves) to measure the dielectric changes at the substrate surface. However, this type of PC structure has not been used for realizing high spatial resolution biosensor imaging since its Bloch surface modes are not confined laterally (rather they propagate along the plane of the substrate surface). Another type of important PC structure is the PC slab, which consists of a periodicity of RI contrast in the plane of the substrate surface introduced by alternating a high-RI guiding layer (e.g., TiO_2, GaAs) with low-RI materials (e.g., air, water, Si) [7,27,60–74]. The PC slabs are typically comprised of 1D (e.g., linear) or 2D (e.g., quadratic and triangular) structures [7,46,51,63,75], and here we focus on the 1D PC slab since it is the simplest to use for PCEM. A PC slab not only supports in-plane guided modes that are confined by the slab completely (which cannot couple to external radiation), but also supports guided-mode resonances (referred to as quasi guided modes or leaky modes) which can couple to the external environment. Therefore, the maximum intensity of the electromagnetic field can be observed both in the high RI layer and in the evanescent part outside of the PC slab. When used in biosensing and bioimaging, the binding events of biomaterials interacting with the evanescent field atop of the PC slab and the associated RI changes can be obtained by detecting the peak wavelength shift (PWS) of guided-mode resonances in the reflection/transmission spectrum. Since the periodicity is in the plane, the lateral propagation of the modes is prohibited in the PC slab biosensor and, therefore, high spatial resolution can be realized in bioimaging.

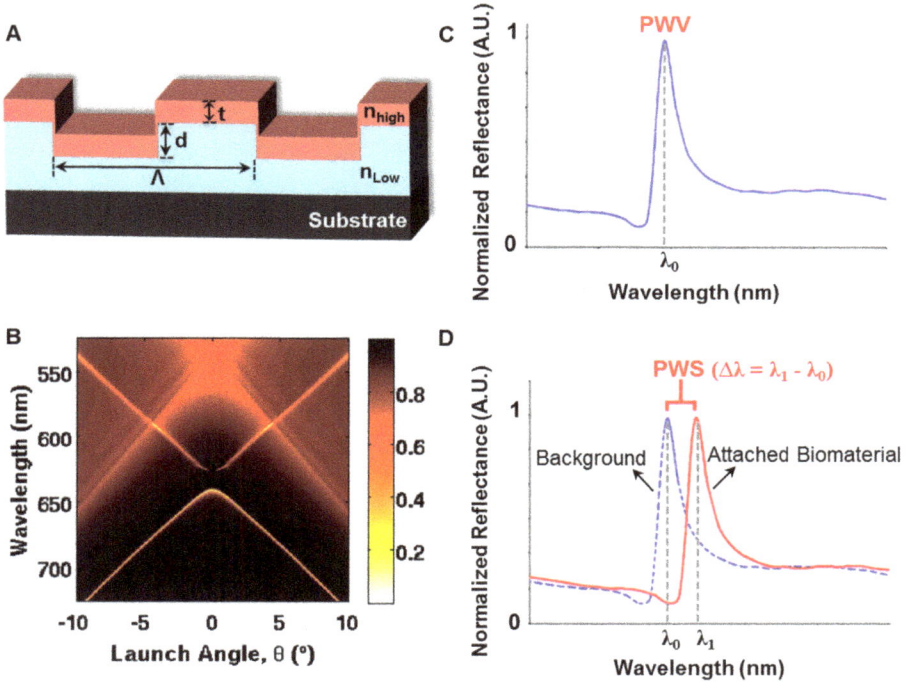

Figure 1. Photonic Crystal (PC) Surface Biosensor. (**A**) Schematic of the PC surface on a substrate with structure parameters: grating period (Λ), grating depth (d), refractive index (RI) of low-RI grating material (n_{low}) and high-RI top layer (n_{high}), thickness of high-RI layer (t); (**B**) Band structure of a photonic crystal biosensor calculated by FDTD simulation; (**C**) Normalized reflection spectrum from the PC surface with resonant peak wavelength value (PWV) of λ_0; (**D**) Peak wavelength shift (PWS) of $\Delta\lambda$ extracted from the normalized spectra with a background pixel (PWV of λ_0) and a pixel with surface-attached biomaterial (PWV of λ_1) on the PC biosensor.

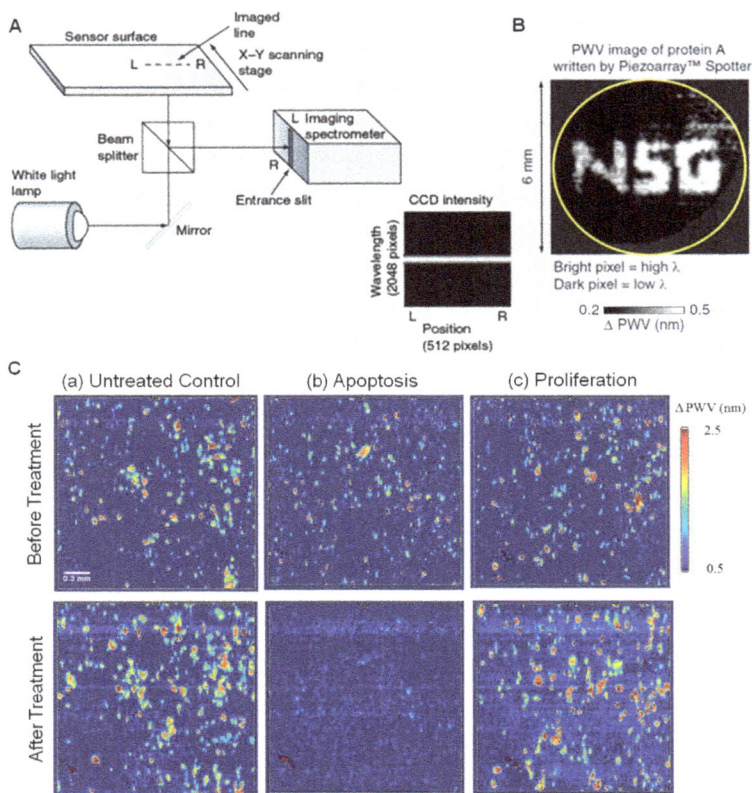

Figure 2. Instrument 1: Label-free Biomolecular Interaction Detection (BIND) Scanner utilizing a PC surface biosensor. (**A**) Schematic of excitation/detection instrument where an imaging spectrometer gathers hundreds of reflected spectra simultaneously from one line across the sensor surface; (**B**) PWS images of Protein A (bright regions represent regions of greater PWS) gathered on a 6-mm diameter region of a PC biosensor, which is imaged at approximately 20 μm pixel resolution after writing the letters 'NSG' (Nano Sensors Group) with a microarray spotting tool (PerkinElmer, Inc. Piezoarray™) (Reprinted in part with permission from [50], © 2006 Future Drugs Ltd.); (**C**) PWS images with shift scale bars (ΔPWV) indicating the magnitude of wavelength shifts in nanometers. Pixels with higher PWS displayed in brighter colors indicate locations where Panc-1 cell attachment has occurred. The three columns of image sets represent the following: (a) untreated control, (b) extract that induced 100% cell death Petunia punctata Paxton (*P. punctate*), (c) extract that enhanced proliferation Anisoptera glabra Kurz (*A. glabra*). The top row of images was taken before exposure and the bottom row of images was taken after the 24 h exposure period with a plant extract at 100 μg/mL. Scale bar (white) = 300 μm (Reprinted in part with permission from [41], © 2010 BioMed Central Ltd.).

34

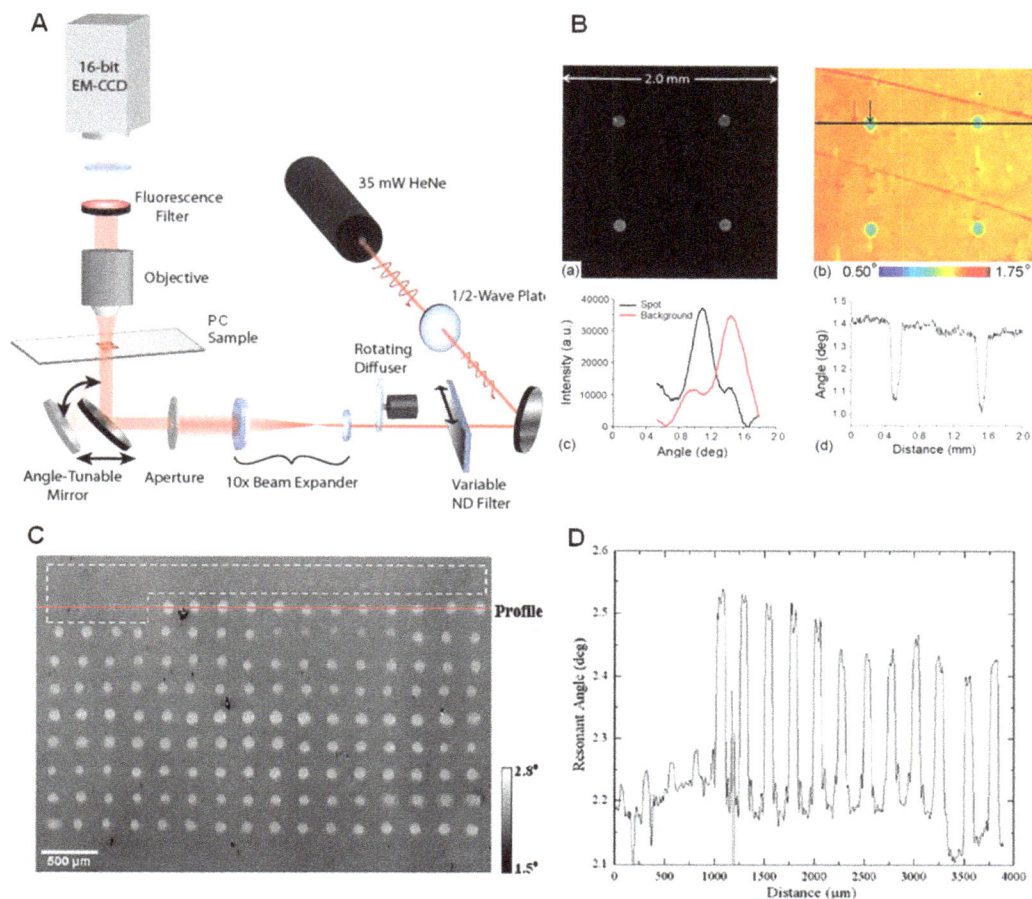

Figure 3. Instrument 2: Transmission acquisition mode of photonic crystal biosensor integrated with an upright imaging microscope and using laser as light source. (**A**) Schematic of combined label-free and enhanced-fluorescence imaging instrument; (**B**) Enhanced (a) fluorescence and (b) label-free images of 50 mg/mL SA-Cy5 spots on a PC biosensor. Inverted transmission *versus* angle response for a pixel inside and outside the SA-Cy5 spot in (c), and cross-section of the label-free image through two SA-Cy5 spots in (d). Rather than measuring the PWS, the label-free imaging system measures the angle of minimum transmission (AMT) by illuminating the PC sensor at a fixed wavelength while scanning the angle of illumination through computer-controlled rotation of the mirror (reprinted in part with permission from [76], © 2009 American Optical Society); (**C**) Label-free image of a DNA microarray measured with a PC biosensor. The white dashed box denotes the location of a set of 20 intentional blank spots. A line profile running through a row containing 4 blank spots followed by 12 probe spots is shown in (**D**) (Reprinted in part with permission from [37], © 2010 American Chemical Society).

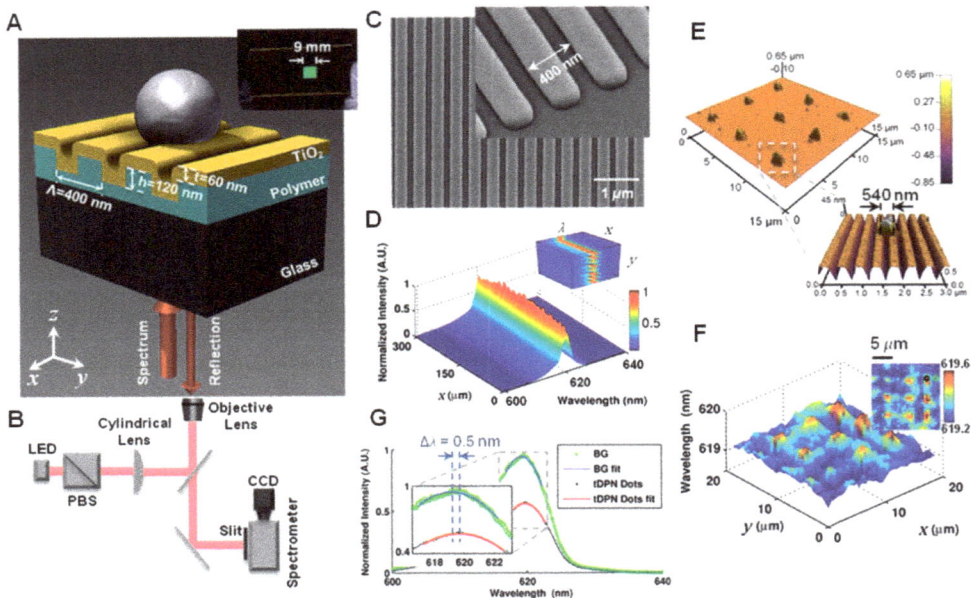

Figure 4. Instrument 3: Reflection acquisition mode of photonic crystal biosensor integrated with an inverted microscope and using LED as light source. (**A**) Schematic of the structure of a photonic crystal (PC) surface biosensor with a surface-attached nanoparticle, inset: photo of a PC biosensor fabricated on a glass slide; (**B**) Instrument schematic of the modern Photonic Crystal Enhanced Microscopy (PCEM); (**C**) Scanning electron micrograph of the photonic crystal surface, inset: zoomed-in image on the edge of the PC biosensor; (**D**) Normalized spectrograph (surface plot) measured with PCEM. Inset: PCEM-acquired 3D spectrum data; (**E**) AFM images of a tDPN-printed 3 × 3 array of nano-dots (each with dimension of $540^2 \times 40$ nm^3), inset: zoomed-in AFM image of one tDPN-printed dot; (**F**) PWV image of the tDPN-printed dots (displayed in a 3D surface plot) within a 20^2 μm^2 field of view, inset: 2D PWV image; (**G**) Normalized spectra of a representative tDPN-printed dot and a background pixel, inset: zoomed-in image of the spectra with 2D polynomial fitting (Reprinted in part with permission from [48], © 2014 RSC Publishing.).

The advantages of PCEM are inherent from the optical properties of slab-based PC surfaces, since they can be designed as a wavelength-selective optical resonator and functionalized as a sensitive optical transducer. For instance, high spatial resolution (in-plane) can be achieved in bioimaging due to the restricted lateral propagation of electromagnetic waves on surface of the PC slab. Enhanced electromagnetic fields (in the form of an evanescent field) near the PC surface (penetration depth of ~200 nm) only illuminate surface-adsorbed biomaterials, such as the extracellular matrix (ECM), membrane components of surface-adsorbed cells, or cellular surface-attached nanoparticle tags. This near-field high-intensity illumination regime promises a high axial resolution (out-of-plane) of ~200 nm, which is beyond the diffraction limit in the spectrum-range of the visible light (400–700 nm) during bioimaging. Compared to the broadband resonances and lossy modes (due to absorption) on metal surfaces, narrow line width (e.g., a few nm) resonant spectra and high reflection efficiency (e.g., nearly 100%) on dielectric surfaces of PC biosensors enable measurement of resonant wavelength shifts with high spectral resolution. The PC resonant mode can be measured in a noncontact detection instrument configuration, in which normal incident-angle illumination results in simple integration with a standard microscope. The resonant wavelength can be selected on a PC surface by tuning its geometry (e.g., grating period) or the incident angle of illumination. Thus, the sensing and imaging can be realized in many spectral ranges, including ultraviolet, visible, and infrared (IR). Although PC surfaces have been fabricated by expensive and time-consuming approaches (such as electron-beam (e-beam) or nano-imprint lithography), recent developments in high-throughput and large-area polymer-based techniques (such as nanoreplica molding at room temperature) have led to the commercial introduction of single-use disposable PC sensors that can be manufactured in a roll-to-roll fashion. These PC sensors can be subsequently integrated with standard format microplates, microscope slides, and microfluidic devices for high-throughput drug or cytotoxicity screening of biomolecule or cell assays. The goal of this review is to summarize the genesis, development, and recent advances of PCEM.

2. Principles of Modern PCEM

2.1. Photonic Crystal Surface Biosensor

A dielectric PC surface (linear PC slab) is utilized as the optical transducer for RI sensing in the label-free PCEM imaging system, as shown in Figure 4A. The PC surface is a resonant grating structure with periodic modulation of the dielectric permittivity of a low-RI material in one dimension (1D) (which provides the nano-pattern) and is then coated with a thin layer of high-RI material (which supports the guided-mode resonances) [49,51]. When illuminated with broadband polarized light, the incident light is coupled into the resonant modes of the PC if the Bragg condition is satisfied. As mentioned earlier, such guided-mode resonances are referred to as "quasi guided modes" or "leaky modes" since they are not allowed to propagate laterally (due to fact that these modes are rapidly re-radiated out from the grating structure) and, thus, have a finite lifetime in the PC structure. Therefore, the resulting electromagnetic standing waves that occur at the resonant wavelength inhibit lateral propagation and open the potential for the PC surface to be utilized for label-free bioimaging. At the combination of incident angle and incident wavelength that satisfies the resonant coupling condition, nearly no light is transmitted through the PC and a high reflection coefficient (~100%) can be achieved during bioimaging [1–3]. The input light can be coupled into the PC resonant mode via wavelength or angle control, which does not require high precision position control and, thus, reduces the complexity of the overall imaging instrument.

Fabrication of the PC surface can be performed upon large-area plastic sheets using a roll-to-roll replica-molding procedure that is performed at room temperature [46,77–79]. The molding template, which can be used repeatedly (up to thousands of times), can be made on silicon wafers or quartz substrates by deep-UV lithography, nano-imprint lithography, or e-beam lithography. During the replica-molding procedure, a thin layer of liquid ultraviolet-curable polymer (UVCP) (low-RI) is deposited on the molding template and then compressed against the device substrate to create a negative volume image of the grating structure from the mold. After exposure to high-intensity UV light, the UVCP is cured to a solid-phase grating structure (e.g., grating period of $\Lambda = 400$ nm, grating depth of $d = 120$ nm, duty cycle of $f = 50\%$). A thin layer of high-RI material (e.g., titanium dioxide (TiO_2)) is subsequently deposited on top of the low-RI grating structure (UVCP), with its thickness (e.g., thickness of $t = 80$ nm) selected to generate a resonant reflection at a specific wavelength (e.g., resonant wavelength of $\lambda_0 = 620$ nm). A scanning electron microscope (SEM) image of a fabricated 1D PC surface is shown in Figure 4C. This replica-molding method provides a rapid, reliable, and inexpensive manufacturing process for PC surface fabrication.

The main criteria for measuring the performance of a PC surface biosensor include sensitivity and spatial-resolution. The sensitivity of a PC biosensor is determined by the material (e.g., the dielectric property of the high-RI layer) or the geometry of the nano-structure (e.g., the thickness of the high-RI layer) [80]. The sensitivity can be estimated with Finite-difference time-domain (FDTD) computer simulations and experimentally characterized with an optical transmission/reflection setup. As mentioned earlier, the spatial resolution of the PC biosensor can be decomposed into in-plane and axial resolution [81]. The in-plane resolution is characterized by the propagation length of resonant modes along the surface plane of the nano-structure and the axial resolution is determined by the penetration depth of the evanescent field atop of the PC surface. In addition, since the PC surface is an optical biosensor, the selectivity is realized by coating the surface-immobilized antibody or ECM molecules on the top of the biosensor. The absence of selectivity constraints on the biosensor avoids the specific design for each application and, thus, enables a broad range of bio-applications for the PC biosensor.

2.2. PCEM Imaging Modality and Operating Principle

As shown in the schematic diagram (Figure 4B), the PCEM detection instrument uses a linear scanning approach and is built upon the body of an inverted microscope (Carl Zeiss Axio Observer Z1). In addition to ordinary brightfield imaging, a second illumination path is provided from a fiber-coupled broadband LED, which is incident on the PC from below. The unpolarized LED output light passes a polarized beam splitter (PBS) to illuminate the PC with light polarized with its axis perpendicular to the grating lines (e.g., y direction), representing the transverse magnetic (TM) mode. Since the resonant wavelength of a 1D PC surface is only sensitive to the incident angle in one angular dimension (perpendicular to the grating) (y direction), the light can be focused in the orthogonal angular dimension (parallel to the grating) (e.g., x direction) to strengthen the incident intensity. Therefore, the light passing through the PBS is focused in one axis (x direction) by a cylindrical lens, while the light remains collimated in the other angular dimension (y direction). The linear beam (collimated direction) is focused on the back focal plane of the objective lens of the microscope. The light emerging from the objective lens (upwards) is thus incident on the PC, so it is collimated in the direction perpendicular to the PC grating lines (y direction) and, thus, all the light reaching the PC with the TM polarization has the same angle of incidence. The reflected light beam passes through the objective lens in the opposite direction (downwards), after which it is projected onto an imaging spectrometer through a narrow entrance slit. The imaging spectrometer contains a diffraction grating that disperses the wavelength components of the PC-reflected light. Once the spectrometer is determined, the dimension of one imaged pixel of the PC in the direction parallel to the grating lines (x direction) is determined by the magnification of the objective lens and the dimension of pixels within the charge-coupled device (CCD) camera (Photometrics Cascade, 512^2 pixels). A motorized stage (Applied Scientific Instruments, MS2000) linearly translates the PC in the perpendicular direction to the grating (y direction). The step-size of the stage (together with the magnifications of the objective lens) determines the pixel size of the PCEM imaging system in the y direction. Therefore, a large area can be scanned in a line-by-line fashion by translating the PC sensor in steps perpendicular to the linear grating direction (y direction). For example, with a 10 × objective lens of the microscope, a 16 μm^2 pixel size of the CCD camera, and a 0.6 μm step size of the motorized stage, a final acquired image with 0.6^2 μm^2 pixel size can be measured in PCEM (with an acquisition speed of ~10 s per frame for a scanning area of 300^2 μm^2).

For PCEM data acquisition, the linear light beam reflected from the PC that contains the resonant biosensing signal produces a spatially resolved spectrum for each point along the line with a narrow bandwidth (e.g., $\Delta\lambda \sim 4$ nm) and forms a 2D spectrograph (e.g., 512^2 pixels) across the line (Figure 4D). After line-by-line scanning, a 3D spectrum data (e.g., 512^3 cube) can be acquired (Figure 4D inset) and the signal/image processing can be performed with computational software (Matlab, MathWorks). Specifically, the spectrum signal can be mathematically fit with a second-order polynomial or Lorentz function for each pixel to extract the peak wavelength and intensity values. With a background image acquired beforehand, shifts in the peak wavelength value (PWV) or shifts in the local peak intensity value (PIV) can be calculated at each pixel location to measure the redistribution of the attached biomaterials.

3. History of PCEM Development

The development of PCEM instrumentation can be described chronologically in three main phases that have led to increasingly finer spatial-resolution and illumination/detection optics, which have been designed for scanning biomolecular layers on dry PCs or cell attachment on PC surfaces exposed to liquid media.

3.1. Instrument 1—Biomolecular Interaction Detection (BIND) Scanner

In 2002, the first PC biosensor introduced by SRU company (SRU Biosystems) was designed for high-throughput microplate-based detection of protein-protein and protein-small molecule interactions, using a PC with resonant reflection in the near-infrared (NIR) spectral range [49,51]. Shortly afterwards, a PC biosensor microplate reader was introduced that incorporated a linear array of optical fibers with illumination/detection heads that could read all the wells in one row (e.g., y direction) of a 96-well microplate at one time [46]. The illumination/detection heads were installed beneath the microtiter plate, which sits upon a motion stage that could translate the plate in an orthogonal dimension (e.g., x direction) to scan the entire microplate in ~15 s. This mode enabled serial re-scanning of the microplate to generate kinetic data for the biomolecular interaction taking place in all the wells. The PC biosensor resonant PWV was determined at each location with this linear scan mode. Subsequently, the first-generation label-free PC biosensor imaging system was introduced and named the "Biomolecular Interaction Detection" (BIND) Scanner (Instrument 1, Figure 2A) [12,47,50]. The optical fiber-based illumination/detection approach was replaced by free space illumination of the bottom surface of the PC biosensor with a broadband light source, and the collection of reflected light into an imaging spectrometer, which was able to rapidly acquire a spatial PWV map by scanning a large sensor surface area. Following the light path of the system, the incoming light beam was divided by a beam splitter, directed to the PC biosensor surface, magnified by an optional objective lens, and, finally, projected into the imaging spectrometer via a narrow entrance slit. The illumination source in this instrument was a white light lamp or a broadband light-emitting diode (LED) in the NIR spectral range, and the detector was a CCD camera. In a single CCD image (Figure 2A, bottom-right inset), the reflectance spectra of several hundred independent locations in one line that spans the PC were gathered at one time. To construct a 2D PWV image, a scanning stage translated the PC across the illumination line in small spatial increments.

The first generation of scanning PC imaging instruments (BIND Scanner) was developed into a commercially available product and utilized in many life science research applications [38–41,43,50,80,82,83]. For example, it has been reported in [83] that assessing combined enhanced fluorescence and label-free biomolecular detection on the same PC surface. The sensitivity of the PC biosensor has been examined in detail in [80]. The PWS image shown in Figure 2B illustrates the detection of a microarray of Protein A printed on the biosensor surface to form the letters 'NSG' (Nano Sensors Group, University of Illinois at Urbana-Champaign) [50]. Cytotoxicity screening of Bangladeshi medicinal plant extracts has been performed with pancreatic cancer cells (Panc-1) using the BIND Scanner. As shown in Figure 2C, the untreated control group and two representative plant extracts, Petunia punctata Paxton (*P. punctate*) and Anisoptera glabra Kurz (*A. glabra*), demonstrate different cellular activities (apoptosis and proliferation) on the biosensor surfaces [41]. The imaging instrument was sufficient for observing large populations of cells with ~9 µm spatial resolution, so that cells with large surface attachment footprints could be observed, although the system lacked sufficient resolution for observing intra-cell attachment dynamics.

3.2. Instrument 2—Transmission Acquisition Mode with Upright Microscopy and Laser Source

To improve spatial resolution, an upright microscope (Olympus BX-51WI) was integrated into the PC imaging system in 2009 [76] and the resulting system was named the "Photonic Crystal Enhanced Microscope" [44]. Instead of measuring reflection efficiency as a function of wavelength from the bottom of the PC surface, the second generation PCEM measured transmission efficiency as a function of incident angle, using a fixed illumination wavelength from a beam-expanded laser (Instrument 2, Figure 3A). This instrument was designed as a wide-field imaging system with collimated angle-tunable laser illumination, which allowed imaging of a PC surface using the same illumination source and imaging optics for both enhanced fluorescence (EF) and label-free (LF) modalities. As shown in Figure 3A, the light beam generated from a HeNe laser passes through a half-wave plate (for polarization control), a variable neutral density filter, a rotating diffuser (to reduce speckle and fringes), a beam expander, an aperture, and a motorized angle-tunable mirror before passing through the PC (which is positioned beneath the microscope objective lens). The gimbal-mounted motorized mirror sits on top of a motorized linear stage in order to maintain a constant illumination area on the PC device (as the mirror rotates) and provide selective light coupling to the PC. Using this approach, high spatial-resolution and high sensitivity LF and EF images (Figure 3B) can be accurately registered with each other since a common beam-path is shared for both imaging modes [76]. An electron-multiplying (EM) CCD camera was used to acquire high resolution and large-area images, and thus enable high-throughput analysis. Moreover, images can be simultaneously acquired with other imaging techniques available on the EF/LF microscope, including reflected brightfield (BF) and differential interference contrast (DIC) images that can be overlaid with EF and LF images.

This transmission-based PC imaging modality that was capable of simultaneous label-free and enhanced fluorescence imaging (EF/LF) was further developed and utilized in several follow-up publications [20,41,42,44,76,81,84]. One of the main applications envisioned for the instrument was for performing DNA and protein microarray analysis, in which the label-free image of immobilized capture spots could be used to verify correct microarray printing and uniform spot density, while the enhanced fluorescence imaging modality would be used after hybridization of the target molecules from a test sample that carries fluorescent tags. Optimization of the imaging spatial resolution was reported in [81]. Microplate, microfluidic channel, and spot-based affinity capture assays were also demonstrated with this detection platform [84]. Figure 3C shows an example of a label-free image acquired with a tunable resonant angle for a DNA microarray immobilized on the biosensor surface [37]. Figure 3D shows a line profile through a row (red line in Figure 3C) containing 4 blank spots followed by 12 probe spots. It can be clearly observed that areas where the probe DNA has been immobilized produce a measurable increase in the resonant angle.

3.3. Instrument 3—Reflection Acquisition Mode with Inverted Microscopy and LED Source

Recently, the PCEM instrumentation transitioned to its third generation, in which an inverted microscope (Carl Zeiss Axio Observer Z1) body was chosen as the base of the system (Instrument 3, Figure 4A,B) [45,48]. While the second generation PCEM was developed specifically for scanning PC surfaces in a dry state for the detection of surface-adsorbed biomolecule patterns (such as DNA microarrays), the third generation PCEM was designed for label-free detection of cells and real-time detection of binding events in which the PC surface is exposed to liquid. In order to avoid scattering and absorption or interference from cell bodies, microfluidic components, semi-opaque liquid media, or liquid-air meniscus, bottom illumination of the PC was adopted in a reflection mode. In this system, detection of resonant reflected wavelength shifts was adopted again as the sensing approach rather than sensing changes in the resonant angle for a fixed illumination wavelength. An LED was chosen as the light source to avoid the speckles in the acquired images that may be caused by a laser illumination source. To obtain higher illumination intensity from the LED light source, a cylindrical lens was added into the illumination light path to convert the incident light from a circular spot to a more concentrated linear beam [45].

Label-free imaging of surface-absorbed live cells (including cell attachment, chemotaxis, and apoptosis) [45] and nanoparticles [48] has been performed using the third generation PCEM. Fluorescence-labeled imaging is also enabled in this system, in which the PC can be excited by a laser illumination source that can couple with the resonant PC mode to obtain an electric field enhancement effect. This enhancement is capable of increasing fluorescence detection sensitivity (which has been validated previously [20,76,83,85–95]) and enabling estimation of the distance of fluorescence emitters from the PC surface [96]. The most recently adopted PC surface design and PCEM detection instrument configurations have already been described in detail in section 2.

4. Applications of PCEM

The PCEM imaging system can be applied to monitor kinetic changes in the spatial distribution of dielectric permittivity for surface-adsorbed materials. This section describes PCEM applications with several examples, such as label-free live cell imaging, nanoparticle and protein-protein binding detection, and intensity enhancement of fluorescent tags embedded within live cells.

4.1. PCEM for Label-Free Live Cell Imaging

Label-free live cell imaging involves a sensing transducer surface, which typically generates an electrical or optical signal when cells interact with it. Biosensors that measure intrinsic cellular properties (such as dielectric permittivity) can be used to determine the number of cells in contact with the transducer, or the distribution/redistribution of focal adhesion areas. Such transducers (e.g., PC biosensors) may be prepared with different surface chemistry coatings that either mimic the *in vivo* microenvironment within tissues or selectively capture specific cell populations through interaction with proteins expressed on their outer cellular membranes. Therefore, the PCEM-based label-free images of cell attachment can assist the study of cell-substrate interactions, including identifying, capturing, and quantifying cells expressing specific surface molecules (Figure 2C) [38–45,50].

Recently, PCEM has been successfully demonstrated as a label-free live cell imaging approach to provide visualizations of each individual cell with subcellular details [45]. As shown in Figure 5A–C, Panc-1 cells were seeded onto a fibronectin-coated PC biosensor and allowed to incubate for 2 h before imaging. The non-uniform distribution of the PWS and the subcellular activity can be visualized clearly for each single cell. Figure 5B shows that the middle cell (No. 2) demonstrates higher PWS in regions near the boundary of lamellipodia formation (consistent with the creation of actin bundles). These darker shadings in the cell indicate regions of higher protein concentration, which may be attributed to higher modulation in the strength of cellular material attachment.

In addition, the kinetics of dynamic interaction between cellular materials and surface coating materials can be measured quantitatively using PCEM. As shown in Figure 5D, a sequence of movie frames demonstrates murine dental stem cells (mHAT9a) gradually attaching on the PC surface. Cells were seeded at 20,000 cells per mL on a fibronectin-coated PC biosensor surface. After three minutes, initial cell attachment appears as small, round regions, which is consistent with spheroid, trypsinized cells coming out of suspension and attaching to a surface. As time progresses, both the size of the cells and intensity of the PWS induced by them increases, indicating a higher localization of cellular material at the biosensor surface, which can be expected during cell spreading. Finally, once cells are sufficiently attached, cellular processes can be observed sensing the cells' microenvironment in all directions. The outer irregular boundaries of the cells have a relatively low PWS (consistent with thin, exploratory filopodia) accompanied by a more heavily attached region slightly immediately adjacent in the cell interior (likely a result of actin bundle formation). Figure 5D illustrates distinct modulation distributions of the attachment strength for both individual cells and the whole cell group during different periods of the adhesion procedure.

4.2. PCEM for Imaging of Nanoparticle and Protein-Protein Binding

Because the PC surface structure restricts lateral propagation of light at the resonant wavelength, it is possible to create spatial maps of the resonant wavelength and the resonant damping that allow high spatial resolution imaging of small-size biomaterials distributed across the surface. Particles smaller than the pixel size (e.g., 600^2 nm^2 for our current PCEM) are very challenging to visualize and identify. However, it is possible to detect the presence of individual particles when the PWS induced by each particle is higher than the detection sensitivity limit of PCEM at each pixel location (the noise-induced PWS need to be considered as well). It is noteworthy that the PWV image for a particle is always within a diffraction-limited distance of up to five (or more) adjacent pixels and, hence, it is not expected to observe a PWS of only one pixel when a sub-micron nanoparticle attaches to the PC. As shown in Figure 4F, a PWV image is acquired for a 3×3 polystyrene particle array that is printed by thermal Dip-Pen Nanolithography (tDPN) [97,98] with heated atomic force microscopy (AFM) tips. Each particle has the dimension of ~$540^2 \times 40$ nm^3 and ~5 μm gaps in between (Figure 4E). Figure 4G demonstrates two acquired spectra (one from a pixel at particle location, and one from background location) and each printed particle can cause ~0.5 nm PWS, which can be easily detected and visualized using the PCEM system. Not only dielectric nanoparticles (as optical scatters) but also metal nanoparticles (as optical absorbers) as small as ~100 nm can be detected via PIV-shift images using PCEM [48].

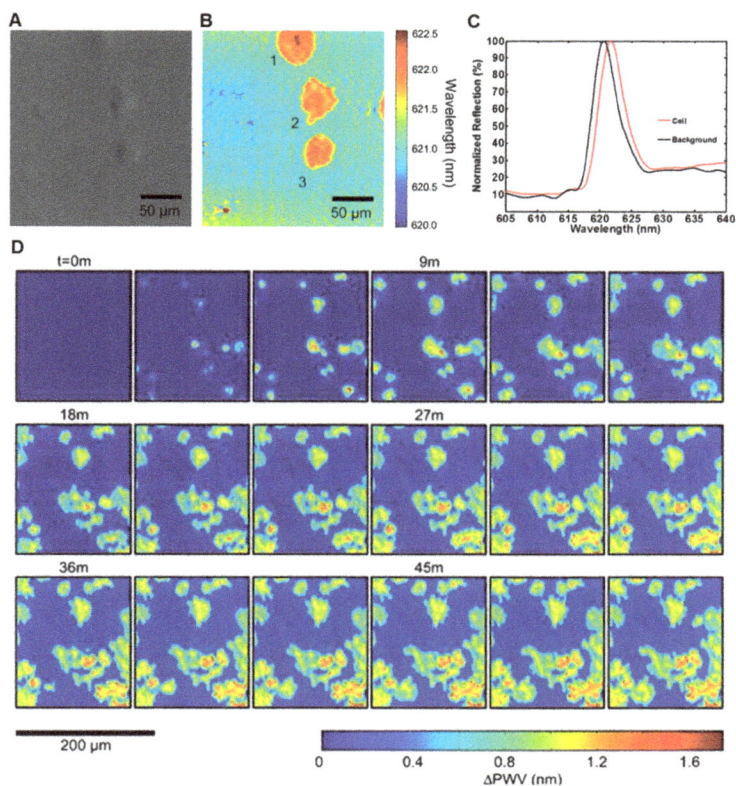

Figure 5. Wavelength-sensitive live cell image from instrument 3–PCEM. (**A**) Brightfield and (**B**) PWV images of Panc-1 cells attached to the PC surface. Lamellipodial extensions are visible, especially from cell 2, demonstrating the ability of PCEM to resolve regional differences in single-cell attachment; (**C**) Representative spectra (normalized) from background regions and regions with cellular attachment. Selected areas of the PWV image from beneath a cell show the PWS of a typical Panc-1 cell is ~1.0 nm; (**D**) Time-lapse PWS images of cellular attachment of dental stem cells (mHAT9a) (Reprinted in part with permission from [45], © 2013 RSC Publishing).

Single nanoparticles allowing direct visualization in PCEM can be applied as biosensing tags to detect protein-protein binding for multiple events on a large sensor surface synchronously. This detection and imaging capability may be used in high-throughput screening during extended periods while avoiding photobleaching issues that are inherent for fluorescence dye tags. Furthermore, the resonance wavelength of nanoparticles can be conveniently tuned through the incident angle of the illumination light [44], the dimension of the PC biosensor [80], and the size or geometry of the nanoparticle [48,99,100]. An example is plotted in Figure 6 for PCEM detection of a target protein molecule (e.g., Rabbit Immunoglobulin G (IgG)) binding with the immobilized capture antibodies (e.g., anti-Rabbit IgG) using gold nanorods (AuNR) as tags [48]. The aspect ratio of the AuNR (dimension of ~$65^2 \times 30$ nm^3) was tuned such that its localized surface plasmon resonance (LSPR) matched the resonant wavelength of the PC biosensor, and thus further improved the signal-to-noise ratio performance of the imaging system.

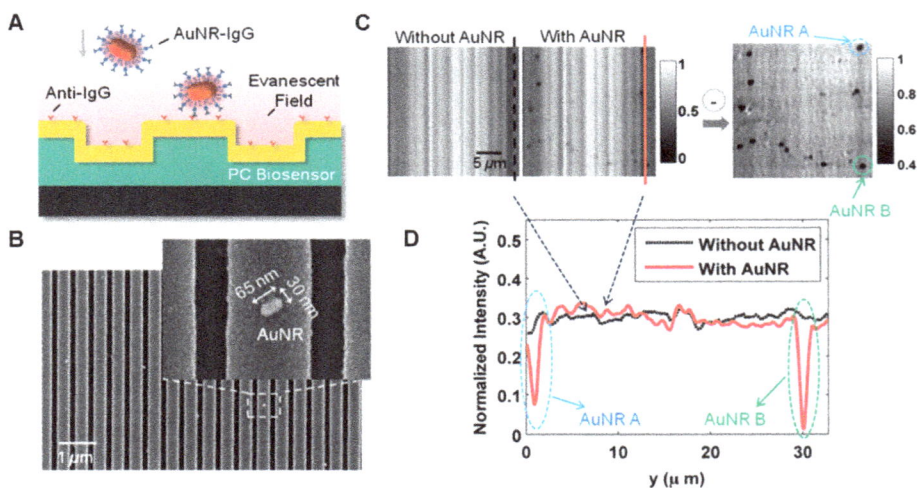

Figure 6. PCEM detection of protein-protein binding. (**A**) Schematic illustration of the PCEM detection of protein-protein binding on the PC biosensor surface; (**B**) SEM images of AuNR-IgG (AuNR conjugated with SH-PEG-IgG) attached to the PC biosensor surface. Inset: zoomed-in image for one AuNR; (**C**) PCEM-detected peak intensity value (PIV) images (in grayscale) and the PIV-shift image indicating AuNR-IgG attached on the PC surface; (**D**) Two representative cross-section lines of the normalized intensity images with/without two AuNRs-IgG on the PC surface (Reprinted in part with permission from [48], © 2014 RSC Publishing.).

4.3. Combination of PCEM and PCEF for Label-Free/Fluorescence-Labeled Imaging Simultaneously

The PCEM is not limited to detection of optical scatters or absorbers, but is also capable of enhancing the emission and extraction from optical emitters (such as fluorescent dyes) in the evanescent field of the PC biosensor. Based on this principle, the label-free PCEM system can be slightly modified to include an additional illumination path from a laser that can excite fluorescent emitters. The ability to tune the illumination angle of the laser to match the resonant coupling condition of the PC substantially enhances the electric field intensity that is used to excite fluorophores, resulting in higher intensity fluorescence microscope images. Photonic Crystal Enhanced Fluorescence (PCEF) represents an additional imaging modality within the PCEM that enables rapid switching between label-free and fluorescence-labeled imaging modes (Figure 3A) [76,83]. Figure 3B demonstrates the enhanced fluorescence image and the label-free image of the same microarray spots printed with cyanine-5-tagged streptavidin (Cy5-SA) proteins. Figure 7A depicts the current optical setup for the PCEF portion of a combined imaging system. Illumination from a fiber-coupled semiconductor laser diode is collimated and passed through a half waveplate to produce a polarization perpendicular to the PC grating lines. Figure 7A inset (top left) plots an angle reflection spectrum of the PC surface when illuminated with a collimated semiconductor laser at 637 nm over a range of illumination angles. Maximum reflection intensity occurs at the on-resonance condition at an incident angle of $\pm1.14°$ from normal direction. The off-resonance condition refers to the laser illumination at an incidence angle of $5°$. Figure 7B illustrates the corresponding enhanced fluorescence images for membrane dye-stained 3T3 fibroblast cells [96]. The combination of both modalities extends the PC-enhanced imaging system to be multi-functional and capable of imaging in numerous bio-applications.

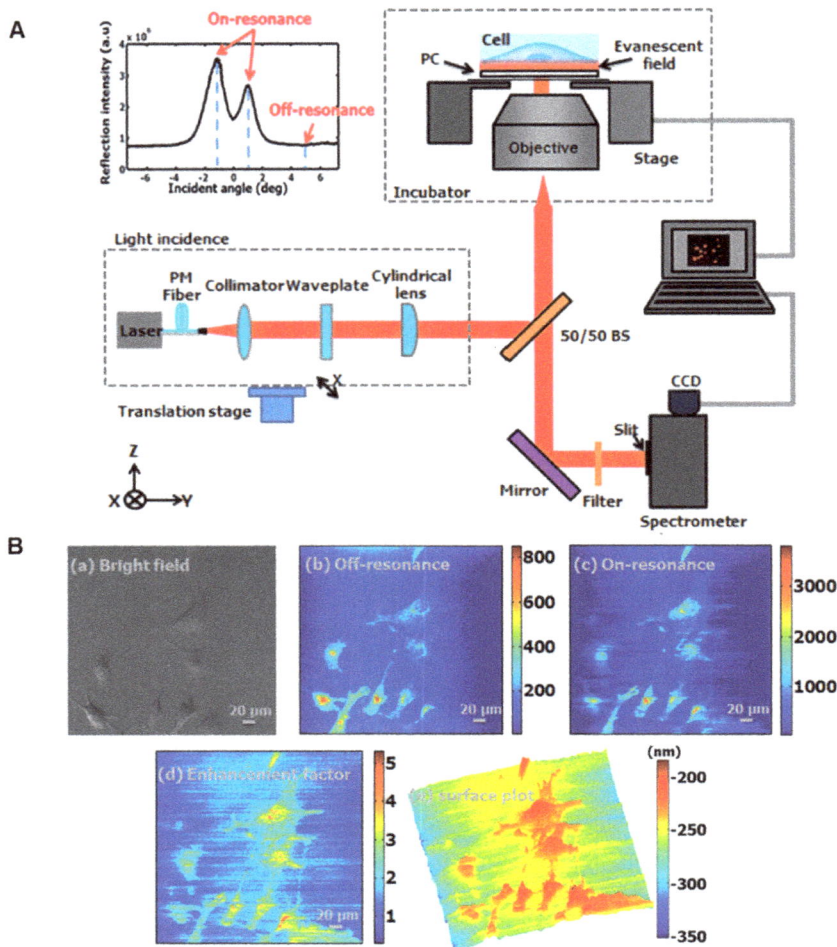

Figure 7. Photonic Crystal Enhanced Fluorescence (PCEF) portion on a PCEM imaging system. (**A**) Schematic of the PCEF portion on modern PCEM detection instrumentation. Inset (top left): angle reflection spectrum; (**B**) Brightfield and PCEF images of membrane dye-stained 3T3 fibroblast cells: (a) brightfield, (b) off-resonance PCEF, (c) on-resonance PCEF, (d) enhancement factor image, (e) 3D surface plot image of the enhancement factor (Reprinted in part with permission from [96], © 2014 RSC Publishing).

5. Summary

Nanophotonic surfaces used in label-free biosensing and bioimaging are an attractive research area and have been involved in many biological applications, including disease diagnostics, drug discovery, and the fundamental study of molecular and cellular activity/function. Detection and imaging tools utilizing nanophotonic surfaces (such as PCEM) with high sensitivity, high detection throughput, and inexpensively manufactured sensors are demanding requirements for life science research and drug discovery applications. This paper reviewed the principles and applications along with the development history of PCEM, which utilizes a photonic crystal surface as an optical transducer to detect and visualize surface-absorbed biomaterials. PCEM achieves high sensitivity and high spatial-resolution due to the narrow spectra line width, restricted lateral propagation and evanescent field enhancement on the PC surface. The PC-enhanced imaging system can be applied to the quantitative and dynamic measurement of cell-substrate interactions, nanoparticle attachment, and protein-protein binding on the biosensor surface. PCEM can also be combined with PCEF to construct a versatile imaging system for tracking and visualizing different optical phenomena that occur within an individual sample. This novel imaging system opens new routes for the detection and visualization of surface-attached biomaterials and holds great potential to help uncover numerous underlying biological mechanisms.

Acknowledgments: This work is supported by the National Science Foundation (NSF Grant CBET 0427657, CBET 0754122, CBET 1132301) and National Institutes of Health (NIH R01 CA118562, R01 GM086382). The authors would like to thank the Nano Sensors Group (NSG) and staff in Micro and Nanotechnology Laboratory (MNTL), the Center for innovative instrumentation Technology (CiiT) at University of Illinois at Urbana-Champaign for their support.

Conflicts of Interest: The authors declare no conflict of interest.

References

1. Hessel, A.; Oliner, A.A. A new theory of wood's anomalies on optical gratings. *Appl. Opt.* **1965**, *4*, 1275–1297.
2. Mashev, L.; Popov, E. Diffraction efficiency anomalies of multicoated dielectric gratings. *Opt. Commun.* **1984**, *51*, 131–136.
3. Popov, E.; Mashev, L.; Maystre, D. Theoretical study of the anomalies of coated dielectric gratings. *Opt. Acta* **1986**, *33*, 607–619.
4. Yablonovitch, E. Inhibited spontaneous emission in solid-state physics and electronics. *Phys. Rev. Lett.* **1987**, *58*, 2059–2062.
5. John, S. Strong localization of photons in certain disordered dielectric superlattices. *Phys. Rev. Lett.* **1987**, *58*, 2486–2489.
6. Joannopoulos, J.D.; Villeneuve, P.R.; Fan, S. Photonic crystals: Putting a new twist on light. *Nature* **1997**, *386*, 143–149.
7. Fan, S.H.; Joannopoulos, J.D. Analysis of guided resonances in photonic crystal slabs. *Phys. Rev. B* **2002**, *65*, 235112.
8. Joannopoulos, J.D.; Johnson, S.G.; Winn, J.N.; Meade, R.D. *Photonic Crystals: Molding the Flow of Light*, 2nd ed.; Princeton University Press: Princeton, NJ, USA, 2008.

9. Magnusson, R.; Wang, S.S. New principle for optical filters. *Appl. Phys. Lett.* **1992**, *61*, 1022–1024.

10. Kikuta, H.; Maegawa, N.; Mizutani, A.; Iwata, K.; Toyota, H. Refractive index sensor with a guided-mode resonant grating filter. *Proc. SPIE* **2001**, *4416*, 219–222.

11. Villa, F.; Regalado, L.E.; Ramos-Mendieta, F.; Gaspar-Armenta, J.; Lopez-Rios, T. Photonic crystal sensor based on surface waves for thin-film characterization. *Opt. Lett.* **2002**, *27*, 646–648.

12. Cunningham, B.T.; Li, P.; Schulz, S.; Lin, B.; Baird, C.; Gerstenmaier, J.; Genick, C.; Wang, F.; Fine, E.; Laing, L. Label-free assays on the bind system. *J. Biomol. Screen.* **2004**, *9*, 481–490.

13. Fang, Y.; Ferrie, A.M.; Fontaine, N.H.; Mauro, J.; Balakrishnan, J. Resonant waveguide grating biosensor for living cell sensing. *Biophys. J.* **2006**, *91*, 1925–1940.

14. Skivesen, N.; Tetu, A.; Kristensen, M.; Kjems, J.; Frandsen, L.H.; Borel, P.I. Photonic-crystal waveguide biosensor. *Opt. Expr.* **2007**, *15*, 3169–3176.

15. Konopsky, V.N.; Alieva, E.V. Photonic crystal surface waves for optical biosensors. *Anal. Chem.* **2007**, *79*, 4729–4735.

16. Nazirizadeh, Y.; Geyer, U.; Lemmer, U.; Gerken, M. Spatially resolved optical characterization of photonic crystal slabs using direct evaluation of photonic modes. In Proceedings of the IEEE International Conference on Optical MEMs and Nanophotonics, Freiburg, Gremany, 11 August 2008; pp. 112–113.

17. Guo, Y.B.; Divin, C.; Myc, A.; Terry, F.L.; Baker, J.R.; Norris, T.B.; Ye, J.Y. Sensitive molecular binding assay using a photonic crystal structure in total internal reflection. *Opt. Expr.* **2008**, *16*, 11741–11749.

18. Fang, Y.; Frutos, A.G.; Verklereen, R. Label-free cell-based assays for gpcr screening. *Comb. Chem. High Throughput Screen.* **2008**, *11*, 357–369.

19. Konopsky, V.N.; Alieva, E.V. Optical biosensors based on photonic crystal surface waves. *Methods Mol. Biol.* **2009**, *503*, 49–64.

20. Cunningham, B.T. Photonic crystal surfaces as a general purpose platform for label-free and fluorescent assays. *JALA Charlottesv Va* **2010**, *15*, 120–135.

21. El Beheiry, M.; Liu, V.; Fan, S.; Levi, O. Sensitivity enhancement in photonic crystal slab biosensors. *Opt. Expr.* **2010**, *18*, 22702–22714.

22. Nazirizadeh, Y.; Bog, U.; Sekula, S.; Mappes, T.; Lemmer, U.; Gerken, M. Low-cost label-free biosensors using photonic crystals embedded between crossed polarizers. *Opt. Expr.* **2010**, *18*, 19120–19128.

23. Jamois, C.; Li, C.; Gerelli, E.; Orobtchouk, R.; Benyattou, T.; Belarouci, A.; Chevolot, Y.; Monnier, V.; Souteyrand, E. New Concepts of Integrated Photonic Biosensors Based on Porous Silicon. In *Biosensors-Emerging Materials and Applications*; Serra, P.A., Ed.; InTech: Rijeka, Croatia, 2011.

24. Magnusson, R.; Wawro, D.; Zimmerman, S.; Ding, Y. Resonant photonic biosensors with polarization-based multiparametric discrimination in each channel. *Sensors* **2011**, *11*, 1476–1488.

25. Nazirizadeh, Y.; Becker, T.; Reverey, J.; Selhuber-Unkel, C.; Rapoport, D.H.; Lemmer, U.; Gerken, M. Photonic crystal slabs for surface contrast enhancement in microscopy of transparent objects. *Opt. Expr.* **2012**, *20*, 14451–14459.

26. Pal, S.; Fauchet, P.M.; Miller, B.L. 1-d and 2-d photonic crystals as optical methods for amplifying biomolecular recognition. *Anal.Chem.* **2012**, *84*, 8900–8908.

27. Threm, D.; Nazirizadeh, Y.; Gerken, M. Photonic crystal biosensors towards on-chip integration. *J. Biophotonics* **2012**, *5*, 601–616.

28. Carbonell, J.; Diaz-Rubio, A.; Torrent, D.; Cervera, F.; Kirleis, M.A.; Pique, A.; Sanchez-Dehesa, J. Radial photonic crystal for detection of frequency and position of radiation sources. *Sci. Rep.* **2012**, *2*, 558.

29. Grepstad, J.O.; Kaspar, P.; Solgaard, O.; Johansen, I.R.; Sudbo, A.S. Photonic-crystal membranes for optical detection of single nano-particles, designed for biosensor application. *Opt. Expr.* **2012**, *20*, 7954–7956.

30. Troia, B.; Paolicelli, A.; Leonardis, F.D.; Passaro, V.M.N. Photonic crystals for optical sensing: A review. In *Advances in Photonic Crystals*; Passaro, V.M.N., Ed.; InTech: Rijeka, Croatia, 2013.

31. Lin, B.; Qiu, J.; Gerstenmeier, J.; Li, P.; Pien, H.; Pepper, J.; Cunningham, B. A label-free optical technique for detecting small molecule interactions. *Biosens. Bioelectron.* **2002**, *17*, 827–834.

32. Chan, L.L.; Cunningham, B.T.; Li, P.Y.; Puff, D. A self-referencing method for microplate label-free photonic-crystal biosensors. *IEEE Sens. J.* **2006**, *6*, 1551–1556.

33. Chan, L.L.; Lidstone, E.A.; Finch, K.E.; Heeres, J.T.; Hergenrother, P.J.; Cunningham, B.T. A method for identifying small molecule aggregators using photonic crystal biosensor microplates. *J. Assoc. Lab. Autom.* **2009**, *14*, 348–359.

34. Ge, C.; Lu, M.; George, S.; Flood, T.A.; Wagner, C.; Zheng, J.; Pokhriyal, A.; Eden, J.G.; Hergenrother, P.J.; Cunningham, B.T. External cavity laser biosensor. *Lab Chip* **2013**, *13*, 1247–1256.

35. Zhang, M.; Peh, J.; Hergenrother, P.J.; Cunningham, B.T. Detection of protein-small molecule binding using a self-referencing external cavity laser biosensor. *J. Am. Chem. Soc.* **2014**, *136*, 5840–5843.

36. Shafiee, H.; Lidstone, E.A.; Jahangir, M.; Inci, F.; Hanhauser, E.; Henrich, T.J.; Kuritzkes, D.R.; Cunningham, B.; Demirci, U. Nanostructured optical photonic crystal biosensor for HIV viral load measurement. *Sci. Rep.* **2014**, *4*, 4116.

37. George, S.; Block, I.D.; Jones, S.I.; Mathias, P.C.; Chaudhery, V.; Vuttipittayamongkol, P.; Wu, H.Y.; Vodkin, L.O.; Cunningham, B.T. Label-free prehybridization DNA microarray imaging using photonic crystals for quantitative spot quality analysis. *Anal. Chem.* **2010**, *82*, 8551–8557.

38. Lin, B.; Li, P.; Cunningham, B.T. A label-free biosensor-based cell attachment assay for characterization of cell surface molecules. *Sens. Actuators B Chem.* **2006**, *114*, 559–564.

39. Chan, L.L.; Gosangari, S.L.; Watkin, K.L.; Cunningham, B.T. A label-free photonic crystal biosensor imaging method for detection of cancer cell cytotoxicity and proliferation. *Apoptosis* **2007**, *12*, 1061–1068.

40. Chan, L.L.; Gosangari, S.L.; Watkin, K.L.; Cunningham, B.T. Label-free imaging of cancer cells using photonic crystal biosensors and application to cytotoxicity screening of a natural compound library. *Sens. Actuators B Chem.* **2008**, *132*, 418–425.

41. George, S.; Bhalerao, S.V.; Lidstone, E.A.; Ahmad, I.S.; Abbasi, A.; Cunningham, B.T.; Watkin, K.L. Cytotoxicity screening of bangladeshi medicinal plant extracts on pancreatic cancer cells. *Complement. Altern. Med.* **2010**, *10*, 52.

42. Shamah, S.M.; Cunningham, B.T. Label-free cell-based assays using photonic crystal optical biosensors. *Analyst* **2011**, *136*, 1090–1102.

43. Chan, L.L.; George, S.; Ahmad, I.; Gosangari, S.L.; Abbasi, A.; Cunningham, B.T.; Watkin, K.L. Cytotoxicity effects of amoorarohituka and chittagonga on breast and pancreatic cancer cells. *Complement. Altern. Med.* **2011**, *10*, 1–8.

44. Lidstone, E.A.; Chaudhery, V.; Kohl, A.; Chan, V.; Wolf-Jensen, T.; Schook, L.B.; Bashir, R.; Cunningham, B.T. Label-free imaging of cell attachment with photonic crystal enhanced microscopy. *Analyst* **2011**, *136*, 3608–3615.

45. Chen, W.L.; Long, K.D.; Lu, M.; Chaudhery, V.; Yu, H.; Choi, J.S.; Polans, J.; Zhuo, Y.; Harley, B.A.C.; Cunningham, B.T. Photonic crystal enhanced microscopy for imaging of live cell adhesion. *Analyst* **2013**, *138*, 5886–5894.

46. Cunningham, B.T.; Qiu, J.; Li, P.; Pepper, J.; Hugh, B. A plastic colorimetric resonant optical biosensor for multiparallel detection of label-free biochemical interactions. *Sens. Actuators B Chem.* **2002**, *85*, 219–226.

47. Li, P.; Lin, B.; Gerstenmaier, J.; Cunningham, B.T. A new method for label-free imaging of biomolecular interactions. *Sens. Actuators B Chem.* **2004**, *99*, 6–13.

48. Zhuo, Y.; Hu, H.; Chen, W.L.; Lu, M.; Tian, L.M.; Yu, H.J.; Long, K.D.; Chow, E.; King, W.P.; Singamaneni, S.; *et al.* Single nanoparticle detection using photonic crystal enhanced microscopy. *Analyst* **2014**, *139*, 1007–1015.

49. Cunningham, B.; Qiu, J.; Li, P.; Lin, B. Enhancing the surface sensitivity of colorimetric resonant optical biosensors. *Sens. Actuators B Chem.* **2002**, *87*, 365–370.

50. Cunningham, B.T.; Laing, L. Microplate-based, label-free detection of biomolecular interactions: Applications in proteomics. *Expert Rev. Proteom.* **2006**, *3*, 271–281.

51. Cunningham, B.T.; Li, P.; Lin, B.; Pepper, J. Colorimetric resonant reflection as a direct biochemical assay technique. *Sens. Actuators B Chem.* **2002**, *81*, 316–328.

52. Yeh, P.; Yariv, A.; Cho, A.Y. Optical surface waves in periodic layered media. *Appl. Phys. Lett.* **1978**, *32*, 104–105.

53. Meade, R.D.; Brommer, K.D.; Rappe, A.M.; Joannopoulos, J.D. Electromagnetic bloch waves at the surface of a photonic crystal. *Phys. Rev. B* **1991**, *44*, 10961–10964.

54. Robertson, W.M.; May, M.S. Surface electromagnetic wave excitation on one-dimensional photonic band gap arrays. *Appl. Phys. Lett.* **1999**, *74*, 1800–1802.

55. Shinn, M.; Robertson, W.M. Surface plasmon-like sensor based on surface electromagnetic waves in a photonic band gap material. *Sens. Actuators B Chem.* **2005**, *105*, 360–364.

56. Descrovi, E.; Frascella, F.; Sciacca, B.; Geobaldo, F.; Dominici, L.; Michelotti, F. Coupling of surface waves in highly defined one-dimensional porous silicon photonic crystals for gas sensing applications. *Appl. Phys. Lett.* **2007**, *91*, 241109-1–241109-3.

57. Sfez, T.; Descrovi, E.; Dominici, L.; Nakagawa, W.; Michelotti, F.; Giorgis, F.; Herzig, H.P. Near-field analysis of surface electromagnetic waves in the bandgap region of a polymeric grating written on a one-dimensional photonic crystal. *Appl. Phys. Lett.* **2008**, *93*, 061108-1-061108-3.

58. Sinibaldi, A.; Danz, N.; Descrovi, E.; Munzertb, P.; Schulz, U.; Sonntag, F.; Dominici, L.; Michelotti, F. Direct comparison of the performance of bloch surface wave and surface plasmon polariton sensors. *Sens. Actuators B Chem.* **2012**, *174*, 292–298.

59. Li, Y.; Yang, T.; Pang, Z.; Du, G.; Song, S.; Han, S. Phase-sensitive bloch surface wave sensor based on variable angle spectroscopic ellipsometry. *Opt. Expr.* **2014**, *22*, 21403–21410.

60. Fan, S.; Villeneuve, P.R.; Joannopoulos, J.D.; Schubert, E.F. High extraction efficiency of spontaneous emission from slabs of photonic crystals. *Phys. Rev. Lett.* **1997**, *18*, 3294–3297.

61. Kanskar, M.; Paddon, P.; Pacradouni, V.; Morin, R.; Busch, A.; Young, J.F.; Johnson, S.R.; MacKenzie, J.; Tiedje, T. Observation of leaky slab modes in an air-bridged semiconductor waveguide with a two-dimensional photonic lattice. *Appl. Phys. Lett.* **1997**, *70*, 1438–1440.

62. Villeneuve, P.R.; Fan, S.; Johnson, S.G.; Joannopoulos, J.D. Three-dimensional photon confinement in photonic crystals of low-dimensional periodicity. *IEEE Proc. Optoelectron.* **1998**, *145*, 384–390.

63. Johnson, S.G.; Fan, S.; Villeneuve, P.R.; Joannopoulos, J.D.; Kolodziejski, L.A. Guided modes in photonic crystal slabs. *Phys. Rev. B* **1999**, *60*, 5751–5758.

64. Painter, O.; Vuckovic, J.; Scherer, A. Defect modes of a two-dimensional photonic crystal in an optically thin dielectric slab. *J. Opt. Soc. Am. B* **1999**, *16*, 275–285.

65. Boroditsky, M.; Vrijen, R.; Krauss, T.F.; Coccioli, R.; Bhat, R.; Yablonovitch, E. Spontaneous emission extraction and purcell enhancement from thin-film 2-d photonic crystals. *Lightwave Technol.* **1999**, *17*, 2096–2112.

66. Astratov, V.N.; Culshaw, I.S.; Stevenson, R.M.; Whittaker, D.M.; Skolnick, M.S.; Krauss, T.F.; de la Rue, R.M. Resonant coupling of near-infrared radiation to photonic band structure waveguides. *Lightwave Technol.* **1999**, *17*, 2050–2057.

67. Baba, T.; Fukaya, N.; Yonekura, J. Observation of light propagation in photonic crystal optical waveguides with bends. *Electron. Lett.* **1999**, *35*, 654–655.

68. Paddon, P.; Young, J.F. Two-dimensional vector-coupled-mode theory for textured planar waveguides. *Phys. Rev. B* **2000**, *61*, 2090–2101.

69. Pacradouni, V.; Mandeville, W.J.; Cowan, A.R.; Paddon, P.; Young, J.F.; Johnson, S.R. Photonic band structure of dielectric membranes periodically textured in two dimensions. *Phys. Rev. B* **2000**, *62*, 4204–4207.

70. Kuchinsky, S.; Allan, D.C.; Borrelli, N.F.; Cotteverte, J.C. 3D localization in a channel waveguide in a photonic crystal with 2d periodicity. *Opt. Commun.* **2000**, *175*, 147–152.

71. Lin, S.Y.; Chow, E.; Johnson, S.G.; Joannopoulos, J.D. Demonstration of highly efficient waveguiding in a photonic crystal slab at the 1.5-um wavelength. *Opt. Lett.* **2000**, *25*, 1297–1299.

72. Benisty, H.; Labilloy, D.; Weisbuch, C.; Smith, C.J.M.; Krauss, T.F.; Cassagne, D.; Beraud, A.; Jouanin, C. Radiation losses of waveguide-based two-dimensional photonic crystals: Positive role of the substrate. *Appl. Phys. Lett.* **2000**, *76*, 532–534.

73. Chutinan, A.; Noda, S. Waveguides and waveguide bends in two-dimensional photonic crystal slabs. *Phys. Rev. B* **2000**, *62*, 4488–4492.

74. Cowan, A.R.; Paddon, P.; Pacradouni, V.; Young, J.F. Resonant scattering and mode coupling in two-dimensional textured planar waveguides. *J. Opt. Soc. Am. A* **2001**, *18*, 1160–1170.

75. Vahala, K. *Optical Microcavities (Advanced Series in Applied Physics)*, 1st ed.; World Scientifc Pubishing Company: Singapore, 2004.

76. Block, I.D.; Mathias, P.C.; Ganesh, N.; Jones, I.D.; Dorvel, B.R.; Chaudhery, V.; Vodkin, L.; Bashir, R.; Cunningham, B.T. A detection instrument for enhanced fluorescence and label-free imaging on photonic crystal surfaces. *Opt. Expr.* **2009**, *17*, 13222–13235.

77. Schulz, S.C. Web based photonic crystal biosensors for drug discovery & diagnostics. *Vac. Coat.* **2008**, 68.

78. Krebs, F.C. Polymer solar cell modules prepared using roll-to-roll methods: Knife-over-edge coating, slot-die coating and screen printing. *Sol. Energy Mater. Sol. Cells* **2009**, *93*, 465–475.

79. Ge, C.; Lu, M.; Jian, X.; Tan, Y.F.; Cunningham, B.T. Large-area organic distributed feedback laser fabricated by nanoreplica molding and horizontal dipping. *Opt. Expr.* **2010**, *18*, 12980–12991.

80. Block, I.D.; Ganesh, N.; Lu, M.; Cunningham, B.T. A sensitivity model for predicting photonic crystal biosensor performance. *IEEE Sens. J.* **2008**, *8*, 274–280.

81. Block, I.D.; Mathias, P.C.; Jones, S.I.; Vodkin, L.O.; Cunningham, B.T. Optimizing the spatial resolution of photonic crystal label-free imaging. *Appl. Opt.* **2009**, *48*, 6567–6574.

82. Choi, C.J.; Cunningham, B.T. Single-step fabrication and characterization of photonic crystal biosensors with polymer microfluidic channels. *Lab Chip* **2006**, *6*, 1373–1380.

83. Mathias, P.C.; Ganesh, N.; Chan, L.L.; Cunningham, B.T. Combined enhanced fluorescence and label-free biomolecular detection with a photonic crystal surface. *Appl. Opt.* **2007**, *46*, 2351–2360.

84. Choi, C.J.; Belobraydich, A.R.; Chan, L.L.; Mathias, P.C.; Cunningham, B.T. Comparison of label-free biosensing in microplate, microfluidic, and spot-based affinity capture assays. *Anal. Biochem.* **2010**, *405*, 1–10.

85. Ganesh, N.; Zhang, W.; Mathias, P.C.; Chow, E.; Soares, J.A.; Malyarchuk, V.; Smith, A.D.; Cunningham, B.T. Enhanced fluorescence emission from quantum dots on a photonic crystal surface. *Nat. Nanotechnol.* **2007**, *2*, 515–520.

86. Ganesh, N.; Block, I.D.; Mathias, P.C.; Zhang, W.; Chow, E.; Malyarchuk, V.; Cunningham, B.T. Leaky-mode assisted fluorescence extraction: Application to fluorescence enhancement biosensors. *Opt. Expr.* **2008**, *16*, 21626–21640.

87. Ganesh, N.; Mathias, P.C.; Zhang, W.; Cunningham, B.T. Distance dependence of fluorescence enhancement from photonic crystal surfaces. *J. Appl. Phys.* **2008**, *103*, 083104.

88. Pokhriyal, A.; Lu, M.; Huang, C.S.; Schulz, S.; Cunningham, B.T. Multicolor fluorescence enhancement from a photonics crystal surface. *Appl. Phys. Lett.* **2010**, *97*, 121108.

89. Pokhriyal, A.; Lu, M.; Chaudhery, V.; Huang, C.S.; Schulz, S.; Cunningham, B.T. Photonic crystal enhanced fluorescence using a quartz substrate to reduce limits of detection. *Opt. Expr.* **2010**, *18*, 24793–24808.

90. Mathias, P.C.; Ganesh, N.; Zhang, W.; Cunningham, B.T. Graded wavelength one-dimensional photonic crystal reveals spectral characteristics of enhanced fluorescence. *J. Appl. Phys* **2008**, *103*, 094320.

91. Mathias, P.C.; Wu, H.Y.; Cunningham, B.T. Employing two distinct photonic crystal resonances to improve fluorescence enhancement. *Appl. Phys. Lett.* **2009**, *95*, 201111.

92. Wu, H.Y.; Zhang, W.; Mathias, P.C.; Cunningham, B.T. Magnification of photonic crystal fluorescence enhancement via tm resonance excitation and te resonance extraction on a dielectric nanorod surface. *Nanotechnology* **2010**, *21*, 125203.

93. Chaudhery, V.; Lu, M.; Pokhriyal, A.; Schulz, S.C.; Cunningham, B.T. Angle-scanning photonic crystal enhanced fluorescence microscopy. *IEEE Sens. J.* **2012**, *12*, 1272–1279.

94. George, S.; Chaudhery, V.; Lu, M.; Takagi, M.; Amro, N.; Pokhriyal, A.; Tan, Y.F.; Ferreira, P.; Cunningham, B.T. Sensitive detection of protein and mirna cancer biomarkers using silicon-based photonic crystals and a resonance coupling laser scanning platform. *Lab Chip* **2013**, *13*, 4053–4064.

95. Pokhriyal, A.; Lu, M.; Ge, C.; Cunningham, B.T. Coupled external cavity photonic crystal enhanced fluorescence. *J. Biophotonics* **2014**, *7*, 332–340.

96. Chen, W.L.; Long, K.D.; Yu, H.J.; Tan, Y.F.; Choi, J.S.; Harley, B.A.; Cunningham, B.T. Enhanced live cell imaging via photonic crystal enhanced fluorescence microscopy. *Analyst* **2014**, *139*, 5954–5963.

97. Hu, H.; Mohseni, P.K.; Pan, L.; Li, X.; Somnath, S.; Felts, J.R.; Shannon, M.A.; King, W.P. Fabrication of arbitrarily-shaped silicon and silicon oxide nanostructures using tip-based nanofabrication. *J. Vac. Sci. Technol. B* **2013**, *31*, 06FJ01.

98. King, W.P.; Bhatia, B.; Felts, J.R.; Kim, H.J.; Kwon, B.; Lee, B.; Somnath, S.; Rosenberger, M. Heated atomic force microscope cantilevers and their applications. *Annu. Rev. Heat Transf.* **2013**, *16*, 287–326.

99. Tian, L.; Chen, E.; Gandra, N.; Abbas, A.; Singamaneni, S. Gold nanorods as plasmonic nanotransducers: Distance-dependent refractive index sensitivity. *Langmuir* **2012**, *28*, 17435–17442.

100. Tian, L.; Morrissey, J.J.; Kattumenu, R.; Gandra, N.; Kharasch, E.D.; Singamaneni, S. Bioplasmonic paper as a platform for detection of kidney cancer biomarkers. *Anal. Chem.* **2012**, *84*, 9928–9934.

Analysis of Surface Plasmon Resonance Curves with a Novel Sigmoid-Asymmetric Fitting Algorithm

Daeho Jang, Geunhyoung Chae and Sehyun Shin

Abstract: The present study introduces a novel curve-fitting algorithm for surface plasmon resonance (SPR) curves using a self-constructed, wedge-shaped beam type angular interrogation SPR spectroscopy technique. Previous fitting approaches such as asymmetric and polynomial equations are still unsatisfactory for analyzing full SPR curves and their use is limited to determining the resonance angle. In the present study, we developed a sigmoid-asymmetric equation that provides excellent curve-fitting for the whole SPR curve over a range of incident angles, including regions of the critical angle and resonance angle. Regardless of the bulk fluid type (*i.e.*, water and air), the present sigmoid-asymmetric fitting exhibited nearly perfect matching with a full SPR curve, whereas the asymmetric and polynomial curve fitting methods did not. Because the present curve-fitting sigmoid-asymmetric equation can determine the critical angle as well as the resonance angle, the undesired effect caused by the bulk fluid refractive index was excluded by subtracting the critical angle from the resonance angle in real time. In conclusion, the proposed sigmoid-asymmetric curve-fitting algorithm for SPR curves is widely applicable to various SPR measurements, while excluding the effect of bulk fluids on the sensing layer.

Reprinted from *Sensors*. Cite as: Jang, D.; Chae, G.; Shin, S. Analysis of Surface Plasmon Resonance Curves with a Novel Sigmoid-Asymmetric Fitting Algorithm. *Sensors* **2015**, *15*, 25385–25398.

1. Introduction

Since the first observation using surface plasmon resonance (SPR) sensors by Wood in 1902 [1,2], SPR sensors have emerged as popular analysis tools for bio-molecules, used label-free to detect changes in the refractive index or thickness of an adsorbed layer on or near the sensing film of the SPR sensor with a high sensitivity in real time [3–9]. However, the performance of the SPR measurement still requires improvement for reliable and high-speed data analysis. In fact, the curve-fitting of the SPR curve is an important and unique process to determine the performance of the SPR sensing, distinguishing the SPR measurement from other direct measurements using cantilever, fluorescence, and electrochemical sensors.

For a typical angular interrogating SPR system, a SPR curve indicating the reflectance intensity *versus* the incident light angle provides a fundamental concept to analyze the binding kinetics of analytes on a sensor film according to changes in the refractive index [10]. SPR sensors generally monitor the changes of reflectance intensity over a range of incident angles when target-molecules interact on the sensing surface. The angle yielding the minimum light intensity on an SPR curve is denoted as the resonance angle, which is carefully determined with curve fitting for an SPR curve in a small range of incident angles. For the accurate measurement of the resonance angle from an SPR curve, several fitting methods have been proposed, such as the polynomial fits [11,12], centroid method [6] and parabolic fit [13,14]. Additionally, optimal linear method [15], asymmetric method [10] and signal processing methods of the SPR signals [16] were proposed. Also to determine the SPR line in the SPR image, researches utilizing Radon transform were introduced [17–21]. In particular, the asymmetric fitting method determines the resonance angle very accurately using a simple equation derived from the complicated multi-layer Fresnel equation.

However, conventional curve-fitting methods have been used for determining the change in the resonance angle in short ranges of the incident angle with wedge-shaped beam type angular interrogation SPR spectroscopy, which is the most popular and appropriate SPR system for real-time monitoring, as shown in Figure 1a. When the targeted molecular interaction is measured by SPR spectroscopy, the real-time results obtained by the change of the resonance angle are also affected by the bulk fluid, which causes a bulk sensor refractive index. In fact, the existence of bulk fluid molecules around a sensing range cannot be avoided and should be excluded in the measured results. Without considering the undesired the effect of bulk fluid molecules, it is difficult to accurately evaluate the net binding kinetics of target molecules by analyzing only the resonance angle.

Conventional SPR devices have adopted a reference channel to remove noise signals caused by the bulky effect of the flowing medium. In order to add a reference channel in a SPR sensor design, it is necessary to give up a main sensing channel on a limited sensor area. Furthermore, noise signals vary greatly with referencing approach [22]. However, these noise signals can be effectively removed by obtaining the critical angle and resonance angle simultaneously without a reference channel when non-specific binding is absent. It is known that the critical angle is related to the refractive index of the surrounding medium [23]. Thus, if the medium is changed, the critical angle would be shifted and the resonance angle also would be shifted, correspondingly. Therefore, the capability to determine both the resonance and critical angles from a SPR curve over an entire range of incident angles is highly required. A successful curve-fitting method for a whole SPR curve can provide both critical and resonance angles. Then, the change of angle on specific adsorption of the analyte can be achieved by subtracting the critical angle from resonance angle in real-time as shown Figure 1b. However, an SPR curve in an entire range of incident angles cannot be easily fitted because of the complicated shape of the curve. The conventional fitting methods are not suitable for fitting the entire SPR curve over a range of incident angles. Although a multi-layer Fresnel equation with curve fitting [24] can determine the critical angle and resonance angle accurately, this has rarely been practically applied because the properties are not fully available and it requires a long computation time [10].

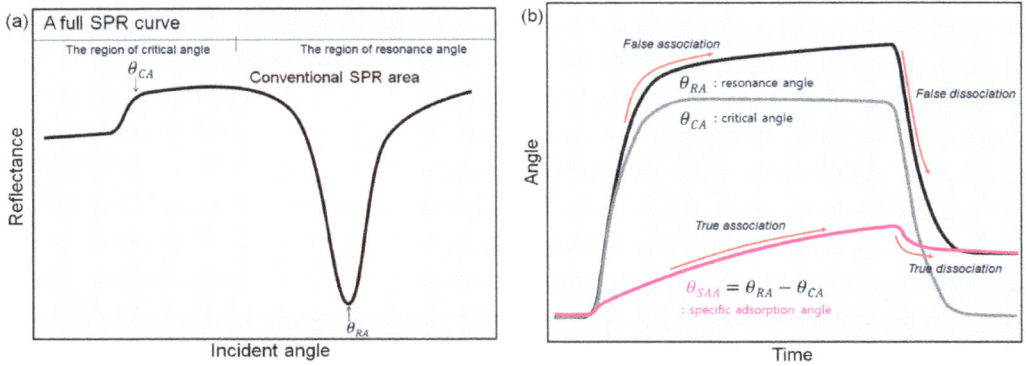

Figure 1. Concept schemes of the proposed sigmoid-asymmetric fitting method: (**a**) A full SPR curve including both regions with critical angle and resonance angle. (**b**) Sensor grams indicating the changes in resonance angle, critical angle and specific adsorption angle, respectively. Our novel method can monitor the change in the specific adsorption angle (θ_{SAA}) by simultaneously monitoring the change in the resonance angle (θ_{RA}) and the change in the critical angle (θ_{CA}).

In this study, we proposed a novel fitting algorithm based on a sigmoid-asymmetric equation that can fit an SPR curve over an entire range of incident angles. The proposed curve-fitting equation is a formulation combining the sigmoid-equation and asymmetric equation. The former equation determines the critical angle, whereas the latter determines resonance angle. Using this curve-fitting method, one can rapidly determine a critical angle as well as a resonance angle. The present analytical results were compared with those for the asymmetric and polynomial equation based fitting methods. In the present study, we also confirmed the feasibility for evaluating of specific adsorption of an analyte on a sensor chip by monitoring in real time the specific adsorption angle subtracting the critical angle from resonance angle with correlation constants.

2. Experimental

2.1. Instrumentation

We fabricated a lab-made wedge-shaped type angular interrogation-based SPR spectroscopy for signal detection. The equipment includes a light source, prism, detector and signal analysis software, liquid handling system with a peristaltic pump and a degasser and the flow cell. A schematic of our angular-interrogation-based Kretchmman-configuration SPR system is presented in Figure 2. A slide glass with a sputtered gold layer (50 nm of Au on 2 nm of Cr) on one side together with the flow cell is pressed against the prism coated with an index matching fluid in order to ensure continuous proceeding of the light. We used a 770 nm light-emitting diode (Opnext Inc., Tokyo, Japan) as a light beam in our system. The p-polarized wedge-type incidence beam with a range of the incident angle of 7.296° (1 pixel = 0.0057°) passes through a band-pass interference filter (770 $\pm$ 10 nm) and is entered to the SPR sensor chip through a half-cylindrical prism. Then, the intensity of the reflected light beam is monitored using a two-dimensional complementary metal oxide semiconductor (2D-CMOS) image sensor (IDS Co., Obersulm, Germany), which has a 1.41 cm sensing area (1280 × 1024 pixels). The image sensor is located immediately in front of the prism, and it allows the SPR system to be fabricated without any other lenses. Our system also has a rotation stage, which can control the incident angle from 35° to 85° as need for the various samples, including gas and liquid solutions. The flow cell is composed of independent three channels with dimensions of 5 mm (l) × 1 mm (w) × 0.2 mm (h) and fabricated from polyether ether ketone (PEEK) plastic. PEEK is used because it has excellent mechanical and chemical resistance properties. The sample solution is driven by peristaltic pump into flow cell and passes through a degasser in order to remove bubbles in the solution before entering the flow cell.

2.2. Image Processing

A final image for the curve fitting is acquired from three images—a dark image, TE-mode image, and TM-mode image—using a self-made MATLAB-based program. The dark image is obtained when the incident light is turned off, while the TE-mode and TM-mode images are obtained from the light-on mode when the polarizer is in the TE-mode and TM-mode, after the running buffer is injected on a gold sensor chip. We processed these three images using the following methods. Firstly, the intensity of the dark image is subtracted from the TE-mode and TM-mode images in order to remove the noise signal in the dark condition. Then, the final image is derived by dividing the subtracted TM-mode image by the subtracted TE-mode image. Figure 3 shows four images obtained from a 2D CMOS image sensor.

$$\text{Final image} = \frac{TM\ mode\ image - Dark\ image}{TE\ mode\ image - Dark\ image}$$

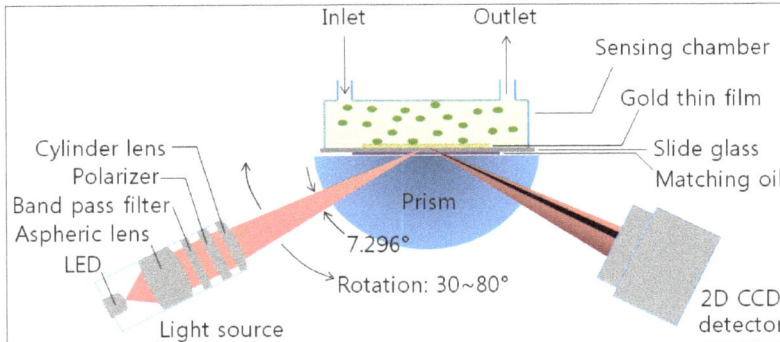

Figure 2. Schematic of our homemade SPR system based on angular interrogation of Kretchmann configuration. The p-polarized wedge-type incidence beam with a range of the incident angle of 7.296° passes through a band-pass interference filter (770 ± 10 nm) and is directed to the SPR sensor chip through a cylindrical prism (BK7, $n = 1.5125$ at 770 nm). Then, the intensity of the reflected light beam is monitored using a two-dimensional complementary metal oxide semiconductor (2D-CMOS) image sensor (IDS Co.) with a 1.41 cm sensing area (1280 × 1024 pixels).

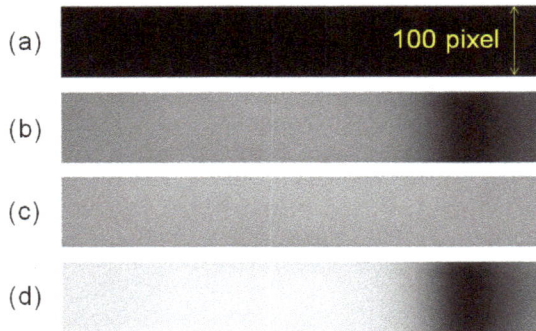

Figure 3. Images from a 2D CMOS image sensor: (**a**) The dark image is obtained when the incident light is turned off. (**b**) The TE-mode. (**c**) TM-mode images are obtained from the light-on mode when the polarizer is in TE-mode and TM-mode after the running buffer is injected on a gold sensor chip. (**d**) The final image is obtained after image processing using (a), (b), and (c).

2.3. Fitting Algorithm Based Sigmoid-Asymmetric Equation

Figure 4 shows the proposed concept of the sigmoid-asymmetric curve-fitting algorithm. A full SPR curve denoted by the black dotted line in Figure 4 is acquired by plotting the average intensity values of 100 rows for each column in the final image, which is processed using the method described in Section 2.2. Then, the full SPR curve is fitted by the proposed sigmoid-asymmetric equation of Equation (1):

$$R\left(X\right) = \left(\frac{A \times [1 - \{B + C \times (X - D)\}]}{(X - D)^2 + E^2}\right) + \left(\frac{F}{1 + e^{G \times (X - H)}}\right) + (I \times X) \quad (1)$$

This equation is a formula combining the asymmetric function Equation (2) [10] with an equation modified from the sigmoid function Equation (3) [25].

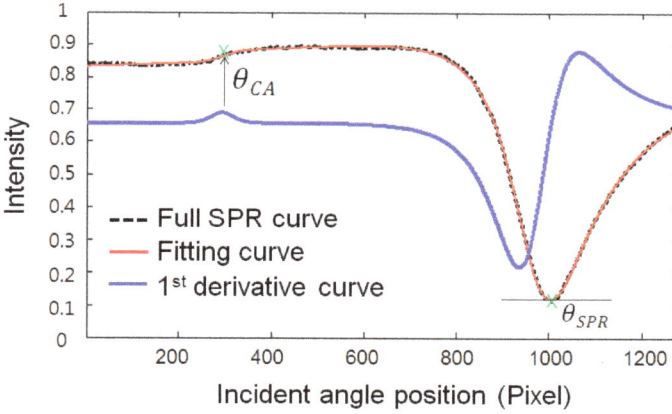

Figure 4. Determination of a resonance angle and a critical angle using sigmoid-asymmetric curve fitting: full SPR curve (black dotted line) plotted from the final image is first fitted by the sigmoid-asymmetric equation (red solid curve), and the resonance angle and critical angle are simultaneously and automatically determined after calculating the 1st derivative (blue solid line).

The asymmetric function contributes to the determination of the optimal resonance angle on the right side of the full SPR curve, and the sigmoid function allows the determination of the optimal critical angle on the left side:

$$R\left(X\right) = \left(\frac{A \times [1 - \{B + C \times (X - D)\}]}{(X - D)^2 + E^2}\right) \quad (2)$$

$$f\left(x\right) = \left(\frac{1}{1 + e^{-x}}\right) \quad (3)$$

In Equation (1), the parameters A, B, C, D, and E are real and constant values needed to fit the right side of the full SPR curve to the asymmetric function, and the parameter X represents the incident angle [11]. The parameters F, G, and H are real and constant values needed to fit the left side of the full SPR curve to the modified sigmoid function, and the parameter I is a real and constant value representing the tilt of the modified sigmoid function. The red solid line of Figure 4 is the fitting curve obtained using the proposed sigmoid-asymmetric equation. The proposed method determines a resonance angle, which is a response angle position to the minimum reflectance on the fitting curve obtained by the sigmoid-asymmetric equation. Moreover, it simultaneously determines the critical angle that is a response angle position to the maximum value of the 1st derivative curve in the region of the critical angle, as indicated by the blue solid line of Figure 4.

2.4. Sample Preparation and Measurements

Chemicals: Glycerol, bovine serum albumin (BSA), phosphate-buffered saline (PBS) were purchased from Sigma, Inc. (St. Louis, MO, USA).

Gold sensor chip: The glass slide (20 mm × 10 mm × 0.55 mm) was from Asahi Glass, Inc. (Tokyo, Japan). The chrome and gold sputtered on the slide with 2 nm and 48 nm of thickness.

Glycerol solutions: Distilled ionized water (DIW) and glycerin solutions of 1%, 2%, 3%, 4% and 5% in DIW were prepared and measured with our SPR instrument to know the relationship between critical angle and resonance angle. First, the DIW was injected into a microchannel on the gold sensor chip for a baseline with flow rate of 40 μL/min. Subsequently, the glycerol-water mixture solutions were loaded at 500 s intervals.

BSA adsorption: BSA of 5 μg/mL in a 1 × PBS with 1.5% glycerin was prepared for protein adsorption in real time to confirm the feasibility of removing the bulk fluid effect. Here, the diluted BSA and glycerin were used as a model protein for adsorption on the gold sensor chip and for artificially changing the bulky refractive index around the sensor film. First, the 1 × PBS was injected into a micro channel on the gold sensor chip for a baseline with flow rate of 40 μL/min. Then, the BSA solution was loaded and then rinsed by 1 × PBS with same flow rate.

3. Results and Discussion

Using a MATLAB-based program developed in-house, we compared the curve fitting results for a SPR curve using three different methods: the asymmetric, 24th-order polynomial regression and sigmoid-asymmetric equations. We excluded the centroid and 2nd order polynomial method, which are also popular methods used in SPR spectroscopy, from our comparison experiments because those are local curve fitting methods with threshold values on SPR curves. In Figures 5 and 6 a full SPR curve for the fitting was obtained by a wedge type angular interrogating SPR sensor system with air and water as bulk fluids on a gold sensor chip, and each fitting curve was plotted, respectively.

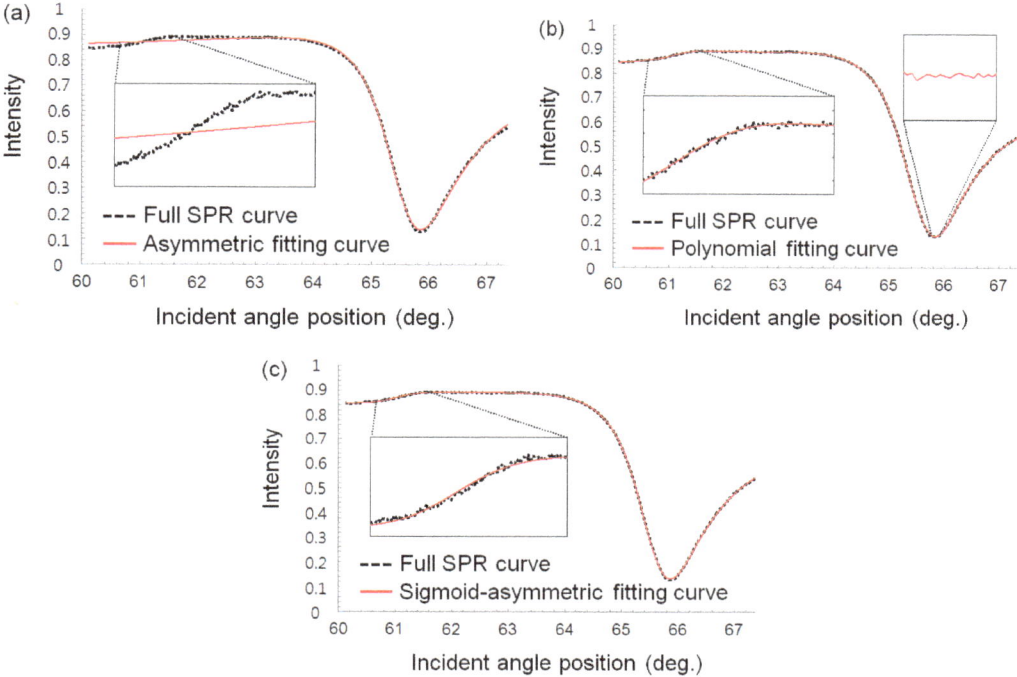

Figure 5. SPR curve-fitting results measured in water (black dot line: SPR curve, red solid line: fitting curve): (**a**) Asymmetric fitted SPR curve. (**b**) 24th-order polynomial regression-fitted SPR curve. (**c**) Sigmoid-asymmetric fitted SPR curve.

Herein, air and water were used to confirm the feasibility in both gas and liquid phases, and with dry and wet samples. Also both fluids are good readily available examples to compare the fit quality according to the curve shapes because the shape of each SPR curve in both bulk fluids is very different. The performance of curve-fitting results with both bulk fluids was compared in the critical angle region and resonance angle region, respectively. The two regions that depict each angle were divided at a criterion angle, which was carefully determined. The criterion angles for water and air were 600 and 550 pixel of the total incident angle, respectively. Available fit quality parameters, including the coefficient of determination (CD), error variance (EV), and angle positions determined by each fitting method, are summarized in Table 1. The coefficient of determination, R^2, indicates how well the experimental data fit the equation models. The EV is used to be defined as follows:

$$\frac{1}{N} \sum_{i=1}^{N} (E_i - \overline{E})^2 \tag{4}$$

where $E_i = X_i - Y_i$ and $\overline{E}$ is average of E_i at each incident angle position. Here, the X_i is the intensity of the fitting curve and Y_i is that of an experimentally obtained SPR curve at each incident angle position i. The resonance angle position was determined at the position where the minimum intensity yields a resonance angle region, whereas the critical angle position was determined at the position yielding the maximum of the 1st derivative SPR curve in the critical angle region.

In Figures 5a and 6a, the red solid lines represent the fitting curves obtained using the asymmetric equation. In the region of the resonance angle (pixel number > 600), the SPR curve was well fitted with the fitting curve. The CDs for water and air was 0.999 and 0.981, respectively and these values are fairly good results compared with sigmoid-asymmetric results (0.999 and 0.997, respectively), as listed in Table 1. However, the region of the critical angle on the left side was not fitted well, as shown Figures 5a and 6a. Consequently, the asymmetric curve-fitting method could not determine the critical angle position except for resonance angle. Thus, the resonance angles determined using asymmetric curve-fitting method were 66.6303° and 42.5929° for water and air, respectively, which agreed well with the sigmoid-asymmetric results (66.6189° and 42.6210°, respectively).

In Figures 5b and 6b, the fitting curves obtained using the 24th-order polynomial regression equation exhibited a very different appearance depending on both bulk fluids. First, the fitting curve for the water condition on the gold sensor chip agreed well with the full SPR curve, but that for air was fitted poorly. The statistical results in Table 1 also clearly indicate the poor curve-fitting for air compared with water. Secondly, even for water, close inspection of the resonance angle region reveals that the fitted SPR curve is not smooth due to the characteristics of the polynomial equation. This unsmooth fitted curve makes determining the minimum resonance angle difficult and subsequently degrades the reproducibility regarding the determination of the resonance angle.

In contrast, the fitting curves obtained using the proposed sigmoid-asymmetric equation almost perfectly matched the whole SPR curve over a range of incident angle, as indicated by Figures 5c and 6c. Immediately after curve-fitting with a sigmoid-asymmetric equation, both the critical and resonance angles could be determined. The determined resonance and critical angles were 66.6189°, 61.8309° and 42.6100°, 41.3208° with water and air, respectively. The determined critical angles almost coincide with theoretical critical angles (61.6265° and 41.3049° with water and air). The quality of the determined angle positions can be verified by the statistical results of curve-fitting in Table 1. The CDs were nearly 1 and the EVs were also relatively small compared with others, regardless of the bulk fluid types.

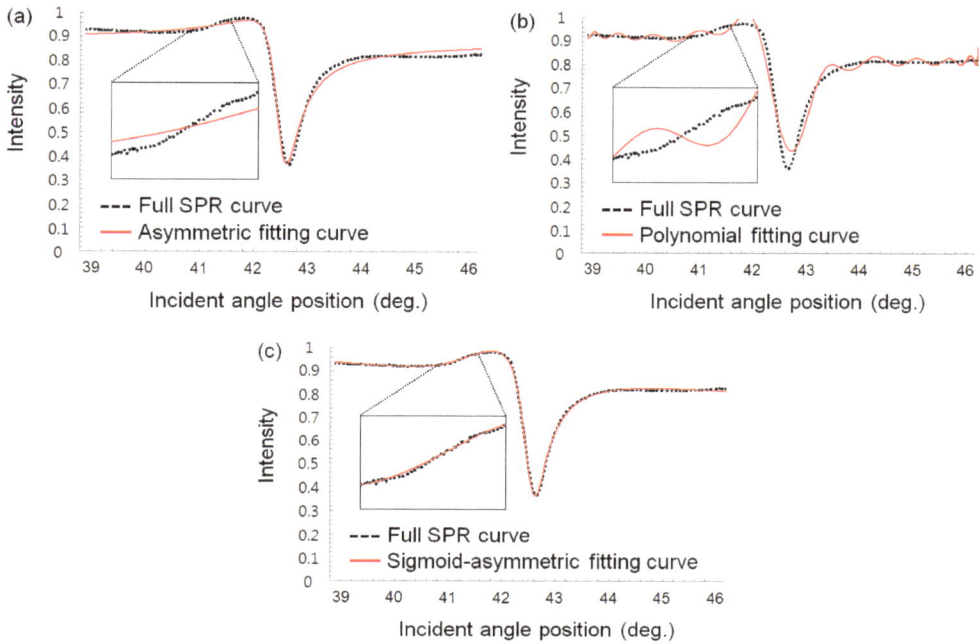

Figure 6. SPR curve-fitting results measured in air (black dot line: SPR curve, red solid line: fitting curve): (**a**) Asymmetric fitted SPR curve. (**b**) 24th-order polynomial regression-fitted SPR curve. (**c**) Sigmoid-asymmetric fitted SPR curve.

As summarized in Table 1, the three fitting methods agreed fairly well in determining resonance angle positions with water as a bulk fluid. The CDs for the three methods are greater than 0.999 and the angle is in the range of 66.6189°–66.6303°. However, for air, the quality of the curve-fitting was generally degraded for all three methods. In particular, the CD and EV for the polynomial method are significantly degraded, and subsequently the corresponding resonance angle (42.6955°) was different from that of other methods (42.5929°, 42.6100°). Meanwhile, the critical angle cannot be determined by any curve-fitting methods except the sigmoid-asymmetric methods, as listed in Table 1. The asymmetric method was unsuitable for determining the critical angle, yielding poor values of the CD and EV for both water and air. Also, the polynomial method was not able to determine the critical angle even though the values of the CD and EV are fairly good for water as a bulk fluid. It is worthy to note that the sigmoid-asymmetric method yielded fairly good fitting results for both water and air. Therefore, the sigmoid-asymmetric method was the only one to fit a whole SPR curve with high quality and thus determine both the resonance and critical angles with precision.

Table 1. Calculated statistical results including error variance and coefficient of determination and both angle positions obtained by fitting methods based on asymmetric, 24th-order polynomial and sigmoid-asymmetric equation for a full SPR curve. The SPR curve was experimentally obtained with water and air as bulk fluids on the sensing film, where *N.A.*: not available.

Bulk Fluid	Region	Fitting Method	Coefficient of Determination	Error Variance $(\times 10^{-4})$	Angle ($^\circ$)
Water	Resonance angle	Asymmetric	0.999	0.268	66.6303
		24th order polynomial	0.999	0.020	66.6303
		Sigmoid-asymmetric	0.999	0.037	66.6189
	Critical angle	Asymmetric	0.536	1.189	*N.A*
		24th order polynomial	0.998	0.004	*N.A*
		Sigmoid-asymmetric	0.995	0.014	61.8309
Air	Resonance angle	Asymmetric	0.981	3.177	42.5929
		24th order polynomial	0.947	8.881	42.6955
		Sigmoid-asymmetric	0.997	0.499	42.6100
	Critical angle	Asymmetric	0.773	0.960	*N.A*
		24th order polynomial	0.412	2.488	*N.A*
		Sigmoid-asymmetric	0.985	0.065	41.3208

In order to monitor specific adsorption of target molecules, one should exclude undesired changes caused by the bulk fluid, which would induce changes in refractive index around the sensor. For this reason, it is necessary to know the relationship between critical angle and resonance angle. The present study monitored the changes in the critical angle and resonance angle on full SPR curve using a DIW and a glycerol-water solution with a concentration in the range of 1% to 5% as a refractive-index solution. In a Figure 7a, the black dotted lines are SPR full curves measured for samples with each concentration of glycerol-water solution and the red solid lines represent fitting curves obtained by sigmoid-asymmetric method. A critical angle and a resonance angle on each curve were determined by the presented algorithm. Figure 7b presents a correlation between the critical angle and resonance angle caused by the change in the fluid refractive index due to the glycerol-water solutions. Fortunately, the correlation represents a simple linear equation in the range of 0.5613°, which is sufficient to measure biomolecular interactions among two or three macromolecular layers in real time as discussed in a previous work [26]. The slope of the trend line in the plots was 0.97, and the coefficient of determination was 0.999. Thus, we determined the normalization constant as 0.97, and the final equation for the specific adsorption angle in our system is described as follows:

$$\theta_{SAA} = \theta_{RA} - 0.97\theta_{CA} \tag{5}$$

where θ_{SAA}, θ_{RA}, and θ_{CA} indicate the specific adsorption angle, resonance angle, and critical angle, respectively.

We conducted additional experiments for protein adsorption in real time to confirm the feasibility of removing the bulk fluid effect using the novel sigmoid-asymmetric equation-based algorithm. The black solid line of Figure 8 represents a sensorgram for measuring the corresponding change in the resonance angle on the BSA adsorption as a measurement method of the conventional SPR system. First, the PBS buffer solution was injected into a micro channel on the gold sensor chip for a baseline. Then, the BSA solution was loaded at the 80 s point. The change in the resonance angle dramatically increased until 200 s. This change reflects a mass increase by the adsorption of the BSA on the surface and the change in the bulky refractive index due to the glycerin concentration. The change in the resonance angle then increased sluggishly until the 700 s point. This change includes only the mass change due to the adsorption of the BSA on the gold surface via hydrophobic interaction. Finally, the change in the resonance angle was dramatically reduced from the 700 s by washing with a PBS buffer solution. The signal was stable at a higher position than the baseline: ~0.0285°. This value indicates the specific adsorption level of the BSA on the gold sensor chip. If we do not know the composition of the sample solution, we cannot understand the meaning of step-by-step changes in the resonance angle.

The sensorgram can be interpreted in many ways. For example, we can attribute the increased signal to the binding of abundant BSA, including strong and weak binding on the gold sensor surface. We can also predict a dramatic decrease in the signal due to the desorption of the weak-binding BSA on the surface. The gray solid line of Figure 8 represents a sensorgram for measuring the change in the critical angle.

We observed that the baseline before loading the BSA-glycerin solution was the same as the last position after the washing with the PBS buffer solution. This sensorgram indicates only the change in the bulky refractive index around the sensor. Finally, we observed a sensorgram to evaluate only the change of specific adsorption angle, *i.e.*, the change in the critical angle caused by the bulky refractive index subtracted from the change in the resonance angle obtained using Equation (5), indicated by the pink solid line in Figure 8. Its value is slowly increased by the adsorption of the BSA before the washing with the PBS buffer, and it is maintained after the washing, not decreasing due to the desorption. We confirmed that the

67

sensorgram is very different from it regarding the change in the resonance angle due to the elimination of the change caused by the bulk refractive index.

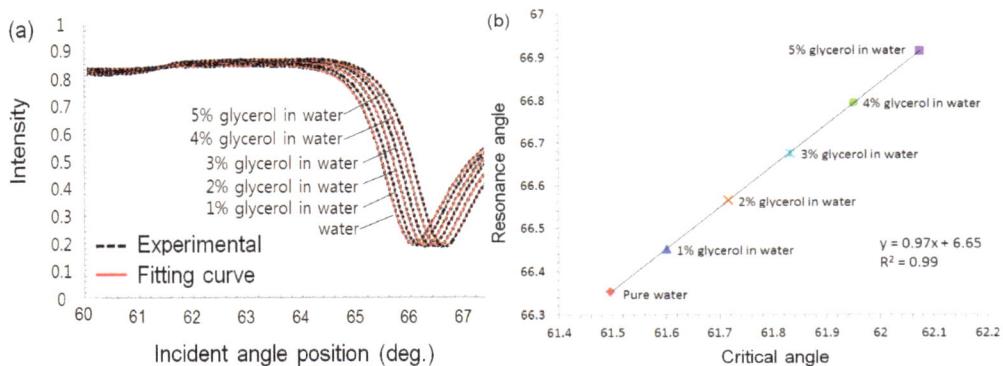

Figure 7. (**a**) Full SPR curves (black dotted lines) measured for a DIW and a glycerol-water solution with a concentration in the range of 1% to 5% as a refractive-index solution and the curves (red solid line) fitted by sigmoid-asymmetric method. (**b**) Correlation graph between the changes in the critical angle and resonance angle caused by the change in the bulky refractive index due to glycerol-water solutions.

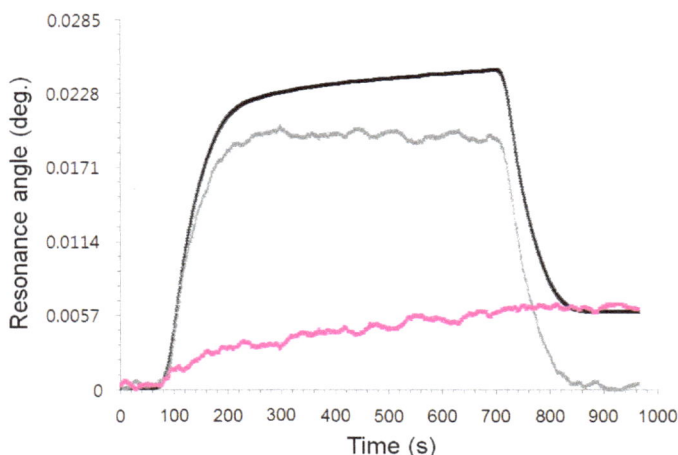

Figure 8. Sensorgrams obtained by measuring θ_{RA} (black solid line), θ_{CA} (gray solid line), and θ_{SAA} (pink solid line) in real time on binding, diluted BSA-glycerin mixture in PBS buffer solution on bare gold sensor chip.

We successfully implemented a sensorgram for measuring the specific adsorption angle by conducting a protein adsorption experiment using only the

68

novel fitting method with a self-constructed wedge-shaped beam type angular interrogation SPR spectroscopy, without any referencing approach or time consuming multi-layer Fresnel equation. We also consider that the sigmoid-asymmetric equation based full SPR curve fitting method is practically useful for the simultaneous and automatic determination of the critical angle and resonance angle in real time.

4. Conclusions

In this study, we introduced a novel full-SPR-curve-fitting algorithm based on a sigmoid-asymmetric equation that can rapidly determine the critical angle and resonance angle in real time. The fitting curves obtained by the proposed sigmoid-asymmetric based approach almost perfectly matched the full SPR curves with water and air as bulk fluids on the sensing film. This was also proven with the available fit quality parameters, which were better than those obtained using fitting methods that are conventionally used to determine the optimal resonance angle, including the error variance and coefficient of determination. The novel algorithm effectively eliminated the undesired change caused by the bulk fluid refractive index change on the sensorgram for measuring the molecular interaction. As a result, we realized a sensorgram for measuring the specific adsorption angle without changes caused by the bulk refractive index, by subtracting the critical angle from the resonance angle in real time using a sigmoid-asymmetric fitting algorithm. We consider that the sigmoid-asymmetric-equation-based full-SPR-curve-fitting method is practically useful for the simultaneous and automatic determination of the critical angle and resonance angle in real time in various applications including gas sensing and solutions based sensing. We believe that the sigmoid-asymmetric fitting equation can be applicable to commercially available SPR systems.

Acknowledgments: This study was supported by the Nano-Material Technology Development Program (NRF-2011-0020090) and Individual Basic Science & Engineering Research Program (NRF-2015R1D1A4A01016653) of the National Research Foundation of Korea, funded by the Ministry of Education, Science and Technology of Korea. This research was also supported by a Korea University Grant.

Author Contributions: All authors contributed to the understanding of the algorithm. Daeho Jang conceived of and designed the algorithm. Geunhyoung Chae performed the simulation analysis and the experiments. Daeho Jang and Sehyun Shin wrote the paper. All authors read and approved the final manuscript.

Conflicts of Interest: The authors declare no conflict of interest.

References

1. Huber, A.; Demartis, S.; Neri, D. The use of biosensor technology for the engineering of antibodies and enzymes. *J. Mol. Recognit.* **1999**, *12*, 198–216.
2. Rabbany, S.Y.; Donner, B.L.; Ligler, F.S. Optical immunosensors. *Crit. Rev. Biomed. Eng.* **1993**, *22*, 307–346.

3. Garland, P.B. Optical evanescent wave methods for the study of biomolecular interactions. *Q. Rev. Biophys.* **1996**, *29*, 91–117.

4. Hutchinson, A.M. Evanescent wave biosensors. *Mol. Biotechnol.* **1995**, *3*, 47–54.

5. Shumaker-Parry, J.S.; Campbell, C.T. Quantitative methods for spatially resolved adsorption/desorption measurements in real time by surface plasmon resonance microscopy. *Anal. Chem.* **2004**, *76*, 907–917.

6. Johansen, K.; Stalberg, R.; Lundstrom, I.; Liedberg, B. Surface plasmon resonance: Instrumental resolution using photo diode arrays. *Meas. Sci. Technol.* **2000**, *11*, 1630–1638.

7. Liedberg, B.; Nylander, C.; Lundström, I. Biosensing with surface plasmon resonance—How it all started. *Biosens. Bioelectron.* **1995**, *10*, i–ix.

8. Wood, R. XLII. On a remarkable case of uneven distribution of light in a diffraction grating spectrum. *Lond. Edinb. Dublin Philos. Mag. J. Sci.* **1902**, *4*, 396–402.

9. Wood, R. XXVII. Diffraction gratings with controlled groove form and abnormal distribution of intensity. *Lond. Edinb. Dublin Philos. Mag. J. Sci.* **1912**, *23*, 310–317.

10. Kurihara, K.; Nakamura, K.; Suzuki, K. Asymmetric SPR sensor response curve-fitting equation for the accurate determination of SPR resonance angle. *Sens. Actuators B Chem.* **2002**, *86*, 49–57.

11. Sjolander, S.; Urbaniczky, C. Integrated Fluid Handling-System for Biomolecular Interaction Analysis. *Anal. Chem.* **1991**, *63*, 2338–2345.

12. Shin, Y.-B.; Kim, H.M.; Jung, Y.; Chung, B.H. A new palm-sized surface plasmon resonance (SPR) biosensor based on modulation of a light source by a rotating mirror. *Sens. Actuators B Chem.* **2010**, *150*, 1–6.

13. Lahav, A.; Auslender, M.; Abdulhalim, I. Sensitivity enhancement of guided-wave surface-plasmon resonance sensors. *Opt. Lett.* **2008**, *33*, 2539–2541.

14. Lahav, A.; Shalabaney, A.; Abdulhalim, I. Surface plasmon sensor with enhanced sensitivity using top nano dielectric layer. *NANOP* **2009**, *3*, 031501–031501-14.

15. Chinowsky, T.M.; Jung, L.S.; Yee, S.S. Optimal linear data analysis for surface plasmon resonance biosensors. *Sens. Actuators B Chem.* **1999**, *54*, 89–97.

16. Karabchevsky, A.; Abdulhalim, I. Techniques for signal analysis in surface plasmon resonance sensors. In *Nanomaterials for Water Management—Signal Amplification for Biosensing from Nanostructures*; Abdulhalim, I., Marks, R.S., Eds.; Pan Stanford: Temasek Boulevard, Singapore, 2015; pp. 163–186.

17. Isaacs, S.; Abdulhalim, I. Long range surface plasmon resonance with ultra-high penetration depth for self-referenced sensing and ultra-low detection limit using diverging beam approach. *Appl. Phys. Lett.* **2015**, *106*.

18. Karabchevsky, A.; Karabchevsky, S.; Abdulhalim, I. Fast surface plasmon resonance imaging sensor using Radon transform. *Sens. Actuators B Chem.* **2011**, *155*, 361–365.

19. Karabchevsky, A.; Karabchevsky, S.; Abdulhalim, I. Nanoprecision algorithm for surface plasmon resonance determination from images with low contrast for improved sensor resolution. *NANOP* **2011**, *5*, 051813–051813-12.

20. Karabchevsky, A.; Tsapovsky, L.; Marks, R.; Abdulhalim, I. Study of Immobilization Procedure on Silver Nanolayers and Detection of Estrone with Diverged Beam Surface Plasmon Resonance (SPR) Imaging. *Biosensors* **2013**, *3*, 157–170.

21. Shalabney, A.; Abdulhalim, I. Sensitivity-enhancement methods for surface plasmon sensors. *Laser Photonics Rev.* **2011**, *5*, 571–606.

22. Springer, T.; Bockova, M.; Homola, J. Label-Free Biosensing in Complex Media: A Referencing Approach. *Anal. Chem.* **2013**, *85*, 5637–5640.

23. Hu, Z.; Cranston, E.D.; Ng, R.; Pelton, R. Tuning Cellulose Nanocrystal Gelation with Polysaccharides and Surfactants. *Langmuir* **2014**, *30*, 2684–2692.

24. Knoll, W. Interfaces and thin films as seen by bound electromagnetic waves. *Annu. Rev. Phys. Chem.* **1998**, *49*, 569–638.

25. Han, J.; Morag, C. The influence of sigmoid function parameters on the speed of backpropagation learning. *Proceedings Series:Lecture Notes in Computer Science* **1995**, *930*, 195–201.

26. Jang, D.; Lim, D.; Chae, G.-H.; Yoo, J. A novel algorithm based on the coefficient of determination of linear regression fitting to automatically find the optimum angle for miniaturized surface plasmon resonance measurement. *Sens. Actuators B Chem.* **2014**, *199*, 488–492.

Considerations on Circuit Design and Data Acquisition of a Portable Surface Plasmon Resonance Biosensing System

Keke Chang, Ruipeng Chen, Shun Wang, Jianwei Li, Xinran Hu, Hao Liang, Baiqiong Cao, Xiaohui Sun, Liuzheng Ma, Juanhua Zhu, Min Jiang and Jiandong Hu

Abstract: The aim of this study was to develop a circuit for an inexpensive portable biosensing system based on surface plasmon resonance spectroscopy. This portable biosensing system designed for field use is characterized by a special structure which consists of a microfluidic cell incorporating a right angle prism functionalized with a biomolecular identification membrane, a laser line generator and a data acquisition circuit board. The data structure, data memory capacity and a line charge-coupled device (CCD) array with a driving circuit for collecting the photoelectric signals are intensively focused on and the high performance analog-to-digital (A/D) converter is comprehensively evaluated. The interface circuit and the photoelectric signal amplifier circuit are first studied to obtain the weak signals from the line CCD array in this experiment. Quantitative measurements for validating the sensitivity of the biosensing system were implemented using ethanol solutions of various concentrations indicated by volume fractions of 5%, 8%, 15%, 20%, 25%, and 30%, respectively, without a biomembrane immobilized on the surface of the SPR sensor. The experiments demonstrated that it is possible to detect a change in the refractive index of an ethanol solution with a sensitivity of 4.99838×10^5 ΔRU/RI in terms of the changes in delta response unit with refractive index using this SPR biosensing system, whereby the theoretical limit of detection of 3.3537×10^{-5} refractive index unit (RIU) and a high linearity at the correlation coefficient of 0.98065. The results obtained from a series of tests confirmed the practicality of this cost-effective portable SPR biosensing system.

Reprinted from *Sensors*. Cite as: Chang, K.; Chen, R.; Wang, S.; Li, J.; Hu, X.; Liang, H.; Cao, B.; Sun, X.; Ma, L.; Zhu, J.; Jiang, M.; Hu, J. Considerations on Circuit Design and Data Acquisition of a Portable Surface Plasmon Resonance Biosensing System. *Sensors* **2015**, *15*, 20511–20523.

1. Introduction

In the last two decades there has been a great effort towards the development of portable surface plasmon resonance (SPR) bioanalyzers to meet the need for fast and non-destructive detection in numerous important areas including food safety, environmental monitoring and agriculture [1–3]. Optical SPR bioanalyzers designed to measure refractive index changes and quantify biomolecular interactions caused by the binding of interacting molecules are typically based on surface plasmons propagating along the metal-dielectric interface where the interaction between an evanescent wave and dielectric occurs [4–6]. However, the price of these bioanalyzers when designed by using a common surface plasmon resonance biosensor is extremely high due to the complicated configurations of the optics and electronics. In recent years, much effort has been dedicated to the development of portable SPR biosensors capable of detecting molecular analytes in real time [7–9]. In practice, portable and cost-effective surface plasmon resonance instruments are urgently needed and have potential in many practical applications, including medical diagnostics, drug screening and basic scientific research. A TiSPR1K23-based biosensor, an integrated SPR biosensor made by Texas Instruments (Dallas, TX, USA), has been used to design a portable bioanalyzer for applications in kinetic analysis of chemical and biological reactions [10–12]. There are a few references on data acquisition circuits for SPR biosensing systems, although the circuit design plays a vital role in the fabrication of bioanalyzers. In this paper we describe a data acquisition circuit for collecting the response signals from a line charge-coupled device (CCD) array and the data transmission from the SPR biosnesing system to the upper computer, mainly composed of a high performance microcontroller, a driving circuit for adjusting the current for the laser generator, a watchdog circuit for monitoring the power supply, and an extension data memory for storing the initialized parameters [13]. A high speed, 12-bit built-in A/D converter is used to collect the signals from the line CCD array. The data acquisition circuit and the corresponding data algorithm to collect the photoelectric signals from the line CCD array were successfully built. The collected photoelectric signals are used to calculate the locations of the surface plasmon resonance dip on the line CCD array in order to perform the association and disassociation processes of biomolecules dynamically [14,15]. The data algorithms are considered extensively to establish the response curve of this SPR biosensing system. Quantitative measurements for validating the sensitivity were implemented in this paper. The outline of the paper is as follows: in Section 2, we briefly review the structure and fundamental principles of SPR biosensing system. Section 3 provides a detailed account of the data acquisition circuit developed for the portable SPR biosensing system, while our experimental results are presented in Section 4. The paper ends with a summary in Section 5.

2. Experimental Section

2.1. Materials

The laser line generator (dimension φ 16 mm × 45 mm, wavelength 780 nm, beam divergent angle 65°) was purchased from SFOLT Co., Ltd. (Shanghai, China). The line CCD array (UPD3575 module) was purchased from Tianjin Brilliance Photoelectric Technology Co., Ltd. (Tianjin, China). A BK7 prism with 50 nm Au film was customized by Changchun Dingxin Photoelectric Co., Ltd. (Changchun, China) The optical adjustment clamp which is designed to hold the right angle prism was fabricated in Henan Nongda Xunjie Measurement and Testing Technology Co., Ltd. (Zhengzhou, China). Ethanol solutions with concentrations of 5%, 8%, 15%, 20%, 25% and 30% volume fraction were purchased from Shanghai General Chemical Reagent Factory (Shanghai, China). Double distilled water was used throughout the whole experiment. 0.01 M PBS (pH 7.4) buffer was prepared by dissolving 0.24 g KH_2PO_4, 8.0 g NaCl, 1.44 g K_2HPO_4 and 0.2 g KCl in 1000 mL of double distilled water.

2.2. Design of the SPR Biosensing System

The prototype of the SPR biosensing system is shown in Figure 1. From the figure, this SPR biosensing system consists of a laser line generator, a microfluidic cell, a line CCD module with driving circuit and an adjustable clamp and a power supply module. In principle, this SPR biosensing system uses a prism, on which surface a 50 nm thick, 1 mm long and 3 mm wide Au thin film was deposited. The dimensions of the microfluidic cell are 3.5 mm (L) × 0.5 mm (W) × 0.25 mm (H). The laser line generator with a P-polarizer is utilized to excite the free electrons which originally are oscillating inside the metal film (Au film). The surface plasmon was produced by the P-polarized laser beam along with the interface between the surface of Au film and biological medium. It is well-known that the evanescent wave produced from the total internal reflection acts on the prism to excite a standing charge density wave on the Au surface [16]. Therefore, a surface plasmon wave will be produced by the standing charge density at the interface between the metal film and the biological medium.

For the biosensor constructed by a prism with the coupling method of the attenuated total reflection, the propagation constants of the incident light wave and the surface plasmon wave along the x axis will be obtained in Equations (1) and (2) (see Figure 1A):

$$K_x^{pr} = \sqrt{\varepsilon_{pr}}\,\frac{\omega}{c}\sin\theta_{pr} \tag{1}$$

$$K_x^{sp} = \sqrt{\frac{\tilde{\varepsilon}_m \cdot \varepsilon_s}{\tilde{\varepsilon}_m + \varepsilon_s}}\,\frac{\omega}{c} \tag{2}$$

74

where the propagation constants for incident light wave and the surface plasmon wave are indicated with K_x^{pr}, K_x^{sp}, respectively. $\tilde{\varepsilon}_m, \varepsilon_{pr}$ are the complex refractive index of the metal film and the refractive index of the prism, respectively. θ_{pr} is the angle formed between the incident light and the normal line of the prism. ε_s is the refractive index of the biological sample flows through the metal film surface. C is the speed of light and ω is the frequency of the surface plasmon wave.

Both propagation constants will be equal, $K_x^{pr} = K_x^{sp}$ when the surface plasmon resonance phenomenon occurs. At the resonance point, the intensity of the incident light is absorbed greatly. The intensity of reflective light is approximately zero [17,18]. By using this relationship, the refractive index of the biological sample bound on the surface of Au film will be calculated. This is seen as a minimum intensity value in the reflection spectra. The position of the minima is indicative of the chemistry on the surface of the SPR sensor. The shift in the minimum value is a measure of the dielectric constant or refractive index changes on the Au surface [19].

In Figure 1, the overall structure of this biosensing platform, which is composed of the laser liner generator, the linear CCD module, the microfluidic cell and the power supply, is shown in Figure 1B. The side view of Figure 1B indicated with Figure 1C shows the position relationship between the laser line generator and the linear CCD array clearly. In this SPR biosensing system (see Figure 1), the laser line generator does not need to be moved to change the angle of the incident beam, so that the laser line generator is exactly fixed by the adjustable clamp. The low cost of the instrument can be developed using this platform.

3. Considerations on Data Acquisition

3.1. Optimization of Interface Circuits

The interfacing system of this SPR biosensing system is a combination of biological sensing membranes and a photoelectrical signals processing circuit. There are four layers in the architecture which were considered to construct this interfacing system. The bottom layer of this interfacing system is dedicated to transducing the refractive indexes changed on the Au film surface of the SPR biosensor into voltage signals (biosensor) in real time, including the linear CCD array and on/off control module of the SPR biosensors. The signal conditioning components including amplifiers for amplifying the photoelectric signals formed the second layer. A microcontroller was used to execute the filtering algorithm to form the third layer. The upper layer, mainly referring to the computer for collecting data from microcontroller with RS232C communication protocol, is used to obtain the response curves and analyze the response unit signals (RUs) [20]. The light intensity of the laser line generator can be controlled with currents through the I/O port of the microcontroller (see Figure 2). For this biosensing system described in Figure 2, once the refractive index is changed with the concentrations of biological samples, these changes will be converted into electronic signals by the SPR biosensor on which the biomolecular identification membrane should be immobilized in advance, while the control process strategy can be implemented by the PIC24FJ128GA008 microcontroller, a very versatile piece of hardware. It has been utilized to receive the data from the A/D converter and the laser line generator is perfectly arranged together to calculate the SPR pixel positions, to plot the SPR curves, to perform the kinetics analysis and to transmit the data to the upper PC. This advanced biological sensing system exists to monitor and process the changes of refractive index efficiently [21]. Obviously, the microcontroller plays an important role in data acquisition and decision implementation.

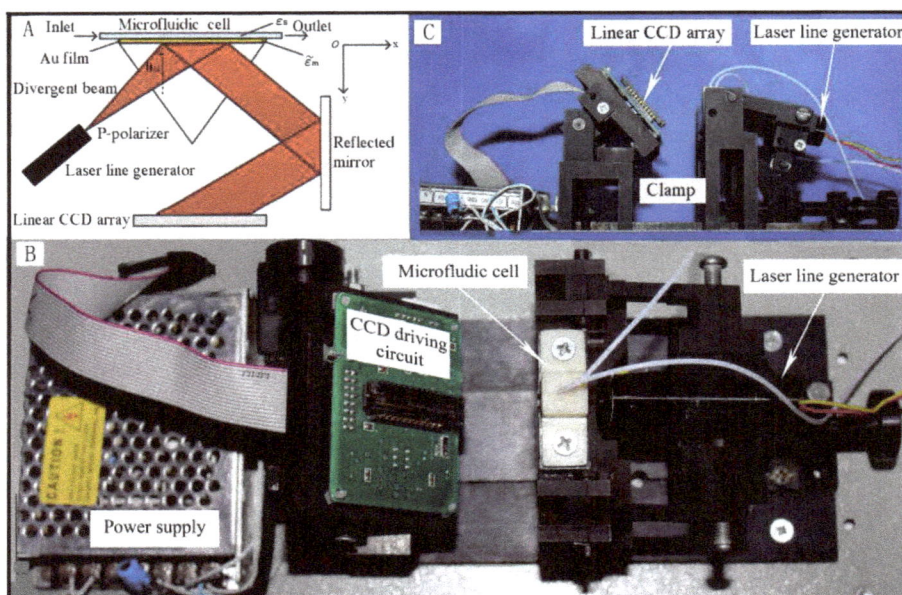

Figure 1. The SPR biosensing platfrom designed by using a laser line generator, a linear CCD module, a microfluidic cell and corresponding clamps. (**A**) Schematic diagram of the principle of this SPR biosensing platform; (**B**) The top view of the overall structure of this SPR biosensing platform without an instrument enclosure; (**C**) The side view of Figure 1B.

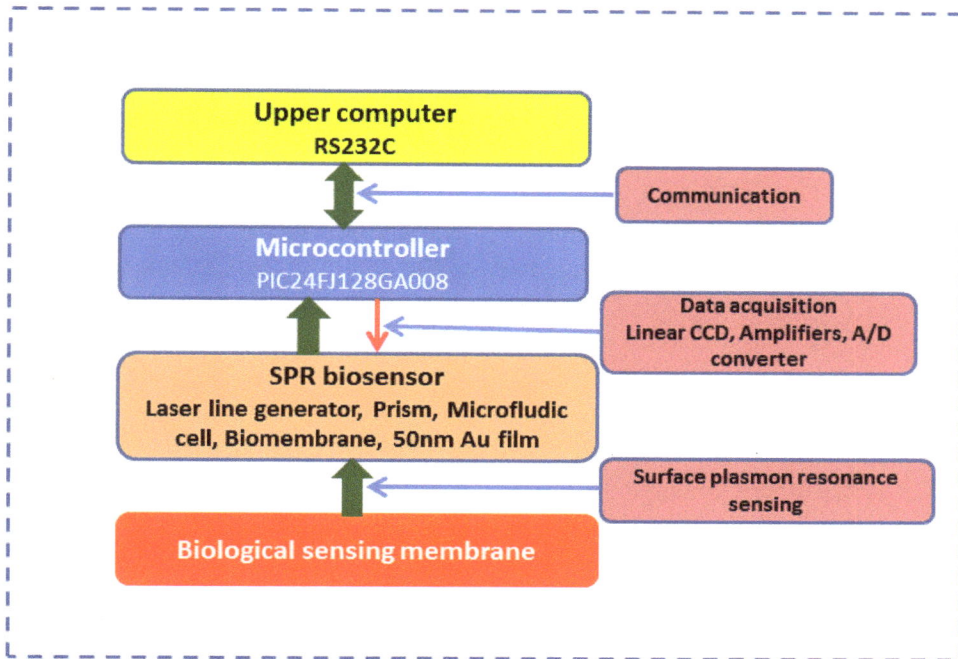

Figure 2. Schematic diagram of the interface system of this SPR biosensing system involving the SPR biosensor, the microcontroller and upper computer.

3.2. Data Structure for Organizing and Storing Response Unit Signals (RUs)

A data structure is considered to organize all the data from the CCD circuit embedded in the SPR biosensing system and from the memory associated with the microcontroller efficiently. The 1024 photoelectric signals from the 16-bit A/D converter are quantified as 16-bit binary codes if the 16-bit A/D converter in the CCD circuit was chosen. The response unit signals (RUs) were computed based on the following formula RU = (1.334 − RIx) × 30,000, where 1.334 is the refractive index of deionized water. RIx is the refractive index of an unknown sample, which can be measured by the SPR biosensing system and 30,000 is a pre-determined factor for increasing the sensitivity of the calculated responses [22]. The normalization of RU values is obtained from the 16-bit A/D converted value from the line CCD array when the biological sample flowed through the Au film surface, which is divided by the 16-bit A/D converted value from the CCD array when air is occurred over the Au film surface [23–25]. Therefore, the RU value is in 16-bit binary code. It is known that the lowest and the highest RU values correspond to 1 and 65,536, respectively. The data structure was intensively considered by taking the least required storage space into account. Certainly, the minimum amount of the required storage space is not only considered in this biosensing system, but also the efficiency for data retrieval should be linked. In this experiment, the RU values only need to be stored as integer type data, such as $-32,768$ and $32,767$. Therefore, the short integer type data structure was chosen due to the fact it only occupies 2 bytes for one measurement value.

3.3. Memory Management

There are no EEPROM units in the PIC24FJ128GA008 microcontroller, therefore, a 24LC256 extension memory was used, which is a 32 K $\times$ 8 (256 K bits) serial electrically erasable PROM. This device also has a page write capability of up to 64 bytes of data to greatly prolong this device's lifetime. A record index and corresponding measurement results are included in each data set. The record index is used to mark the location of the measurement results stored in the extension memory, which is a type of nonvolatile memory. In this experiment, the data for calculating the RU values, for finding the internal data record, and for the communication with the upper computer are stored in this extension memory due to the limitations of the internal memory in the microcontroller [26,27].

The memory capacity of this extension memory is suggested to be 32 KB (32,768 Bytes). If the RU value is expressed in a long integer form it can be used to store a maximum of up to 32,768/4 (8192) measurement results. Three different areas need to be defined in the extension memory device, which are involved to the memory space of record indexes, measurement results and parameters for performing the biosensor actions [28]. The parameters for running the SPR biosensing system are stored in the parameter areas which use a reserved space of 16 or 32 bytes in the extension memory. The sequence numbers (serial number), data status (valid/invalid), channel numbers, the first address of this extension memory, total data capacity and the corresponding measurement information are stored in the record index area. The measurement results are stored in the corresponding format of the data set in the data area [29]. Generally, the block 0 area is used to store the parameters and record indexes, while the other remaining spaces in the extension memory are allocated to store measurement results. In this experiment, block 0 was used to store the parameters and the indexes of the record index area. The record indexes are stored in the area of block 1 and the measurement results are stored in the following blocks in order to upgrade the memory capacity easily. In the block 0 area, the first 128 bytes in this memory are used to store the parameters, while the following 128 bytes are used to store the indexes of the record index area. In the last half part of the block 0, the first 16 bytes of this space were used to store the indexes of the sequence number's index and the first address of the measurement results, while the next 32 bytes of this space are reserved to store the index of the sequence number of the measurement results' index. In the following record index section of the last half part of block 0, there are 8 bytes occupied by each record index. These are the serial number (1 byte) produced by the record indexes, valid status (1 byte), the number of channels (1 byte), the starting address (2 bytes), the total spaces used (2 bytes), and 1 byte is reserved. In the record index of block 1, the record indexes include the measurement results record number, valid status, number of channels, starting address and total of the data, *etc.* In the data area, one data section can store 64 measurement result records if the measurement results are expressed in a long integer form [30]. The memory allocations of the 25LC256 memory chip are shown in Table 1.

Table 1. Memory allocations in the 25LC256 memory chip.

Block 0	Parameters area and the index of the record index area
Block 1	Measurement result record Index area, which indicates the measurement result record number, valid status, number of channels, starting address, total of the data, *etc.*
Block 2	Measurement results area
. . . . . .	. . . . . .
Block 127	Measurement results area

3.4. The Parameter Settings of the Circuit Module of UPD3575D

The photoelectric sensor, linear CCD containing grids of pixels, characterized by photoelectric conversion, charge storage and charge transfer. The output voltage is proportional to the charge packets which are collected in potential wells created by applying a positive voltage to the gate electrodes [31]. Applying a positive voltage to the gate electrode in the correct sequence transfers the charge packets. In this experiment, the photoelectric signals from the line CCD array in the UPD3575 module are obtained under the timing diagrams. The output voltage (V_{out}) of the pixel signals starts to change when the arrival of the falling edge of the pixel synchronizing pulse (PSP) is coming [32].

From Figure 3, the relationship between the V_{out} and the pixel synchronizing pulse is illustrated by the fact that the V_{out} is kept changing when the level of the pixel synchronizing pulse drops to a low value until the coming of the rising edge of the pixel synchronizing pulse, and a stable V_{out} will be achieved when the level of the pixel synchronizing pulse becomes high in the first half cycle. Then the V_{out} will become zero at the high level in the last half cycle of pixel synchronizing pulse [33,34]. The cycle settings of the pixel synchronous pulse are determined manually. The pixel synchronizing pulse's cycle can be set in 2 µs, 4 µs, and 8 µs, electronically.

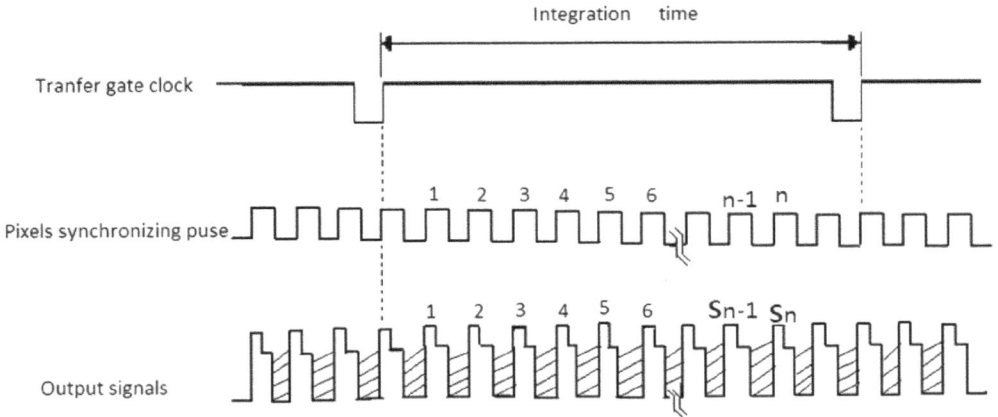

Figure 3. Timing diagram for the photoelectric signals acquisition of the line CCD array.

3.5. Considerations of A/D Converter

The maximum conversion rate is only 100 K samples per second (SPS), so the A/D converter ADS8320 is primarily considered. The 10 µs sampling period will be calculated from the conversion rate of 100 K SPS, which is larger than the maximum pixel synchronizing pulse's cycle of 8 µs obtained from the UPD3575D module. Therefore, a high speed AD converter must be chosen. For the consideration of the microcontroller PIC24HJ32GP302, an A/D converter with a high conversion rate of 1.1 MSPS@10bit (MSPS, million samples per second) or 0.5 MSPS@12bit embedded inside it is chosen. If high AD resolution is needed, the 12-bit resolution is the top priority to be chosen. Therefore, the sampling period of 2 µs at the sampling rate of 0.5 MSPS is calculated. Correspondingly, the pixel synchronizing pulse's cycle of 2, 4, and 8 µs, respectively, is appropriate to match this A/D converter. From the description of the data sheet of the PIC24HJ32GP302 microcontroller, in the 0.5 MSPS@12bit mode, the minimum analog-to-digital converter (ADC) clock period (TAD) of 117.6 ns is found. The conversion time (tCONV) is 14TAD (14 × 117.6 = 1646.4 ns) is calculated according to the TAD value. The clock frequency of this microcontroller is found to be 4 × 7.3728 MHz and the clock cycle is 33.9 ns. Therefore, the instruction cycle TCY (TCY, instruction cycle) is calculated to be 2 × 33.9 = 67.8 ns (TCY is less than TAD). 2TCY (2 × 67.8 = 135.6 ns) is more than TAD (Min). It is known that the sampling cycle should be larger than 3TAD (since 3TAD is the minimum ADC sample time) compared with the ADC clock period; under these conditions, the sampling cycle is 18 TAD (18 × 117.6 ns = 2116 ns = 2.12 µs, with the maximum sampling frequency of 0.5 MHz, and the minimum sampling cycle of 2.0 µs, therefore, the 18 TAD meets the hardware requirements). Correspondingly, the sampling frequency is 472 KSPS. Due to the sampling cycle being over 2 μs, pixel synchronizing pulse cycles of 4, 8 μs are chosen, respectively. Compared with the PIC24FJ128GA008 microcontroller, the PIC24HJ32GP302 microcontroller can fit well the actual requirements of this SPR biosensing system because the built-in A/D converter can work at a high conversion rate of 0.5 MSPS@12bit with the minimum ADC clock period (TAD) of 117.6 ns or work at a high conversion rate of 1.1 MSPS@10bit with the minimum ADC clock period (TAD) of 76 ns.

4. Results and Analysis

4.1. Response of the Biosensing System to Ethanol Concentrations

By using ethanol solutions as the standard detected sample, the responses indicated with RU from the biosensing system for the different ethanol concentrations were measured within a concentration range from 5% to 30% (volume fraction) and deionized water was flowed successively over the Au surface of biosensor to obtain the baseline signals. The Au surface of the biosensor was not modified with a ligand as the specific acceptor for capturing the ethanol molecules to avoid influencing the signals produced by the biomolecular identification membrane. The experiments were performed at 37 °C. The obtained calibration curves are shown in Figure 4.

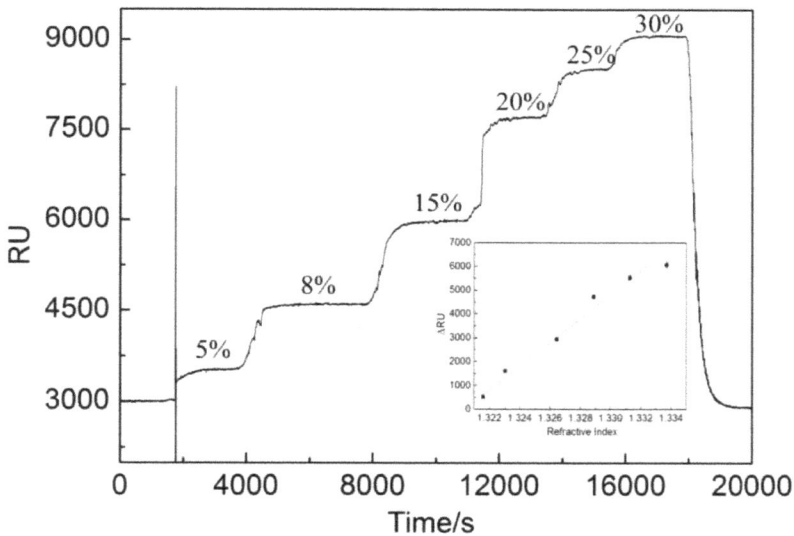

Figure 4. Sensorgrams with inset calibration curve diagrams obtained for different ethanol solution concentrations. The sensorgram was obtained from concentrations of 5%, 8%, 15%, 20%, 25% and 30% ethanol in volume fraction, respectively. The lower right inset indicates the fitting curve established by delta response units with different standard ethanol concentrations ranging from 5% to 30%.

81

The calibration curves represent the process of SPR response signals without a ligand in the dynamic response range of 5% to 30%, which is useful for quantifying purposes. In this experiment, the ethanol molecules started to be adsorbed on the Au film after 1800 s for the reaction in solution. Then the response signals increased rapidly up to a plateau [35,36]. From Figure 4, it is indicated that the association curve gradually but obviously dwindled with increasing ethanol concentration. The plateau of the curve corresponds to the saturation of the sensor active points. A linear range between 5% and 30% can be used for determination of ethanol concentrations.

4.2. Sensitivity Evaluation

The samples with concentrations of 5%, 8%, 15%, 20%, 25% and 30% in volumetric fractions, which can also be converted to refractive indexes of 1.32159, 1.32304, 1.32644, 1.32886, 1.33128 and 1.33370, respectively, were measured five times, repeatedly. The mean response values of these known concentration samples were calculated to be 529, 1607, 2944, 4720, 5541, and 6065 in delta response units, which refers to the sensor response induced by biomolecular binding, changing the local reflective index (RI) at the sensor interface [37]. Importantly, a response (background response) will also be generated if there is a difference in the refractive indices of the running and sample buffers. This background response must be subtracted from the sensorgram to obtain the actual binding response (delta response units, delta RU). Hence, the refractive index of the medium is directly related to the delta RU. The coefficient of variation of the repeated measurement was also calculated to be 5.89%. The fitting equation $\Delta RU = 499837.79883RI-659968.315329$ can be obtained with the R-Square of 0.98065, the theoretical limit of detection of 3.3537×10^{-5} RIU (refractive index unit) and the sensitivity of this SPR biosensing system was calculated to be 4.99838×10^5 $\Delta RU/RI$ (see the inset in Figure 4).

5. Conclusions

The circuit and signal conditioning approaches designed for an inexpensive portable SPR biosensing system constructed using a laser line generator and a linear UPD3575D CCD module have been thoroughly considered. The system is capable of detecting chemical and biological substances and performing kinetic analysis of high affinity biomolecular interactions. The circuit for collecting the signals from the linear CCD array and transferring the measurement results to the computer is mainly composed of a PIC24FJ128GA008 microcontroller, a driver circuit for running the laser line generator, and an extension memory for storing the initialized parameters and measurement results. A UPD3575D CCD module with a 1024 bit linear image sensor capable of converting light into voltage has been chosen and the integration time and the pixel synchronizing pulse's cycle have been discussed in this paper. In this experiment, a high speed, 12 bit built-in A/D converter has been chosen to collect the signals from the linear CCD array. Ethanol solutions with concentrations of 5%, 8%, 15%, 20%, 25% and 30% in volume fraction, respectively, have been used to evaluate the performance of the SPR biosensing system. The ethanol solutions with different concentration factors were flowed over the surface of the sensor chip and the SPR curve and kinetics response curve are established. The measured results for the responses to ethanol showed that the selectivity, detection range, and measuring time of this SPR biosensor supported the utility of the bioassay platform, especially, for low concentration measurements. The experiments demonstrated that it is able to detect a change in the refractive index of an ethanol solution with a sensitivity of $4.99,838 \times 10^5$ $\Delta RU/RI$ in terms of the changes in delta response unit with refractive index, and a high linearity with a correlation coefficient of 0.98065. The theoretical limit of detection of this SPR biosensing system was calculated to be 3.3537×10^{-5} RIU (refractive index unit). Future work will involve the continuation of laboratory tests as well as field trials to obtain more abundant data illustrating the high sensitivity and reliability of this inexpensive portable SPR biosensing system to optimize the algorithm for obtaining the precise position of the resonant dip and the optimization of the circuit design with microcontrollers.

Acknowledgments: This work was financially supported by Henan Province Joint Funding Program (U1304305) of the National Natural Science Foundation of China and State Key Laboratory of Wheat and Maize Crop Science (SKL2014ZH-06) and also supported by Henan Province Science and Technology Cooperation Program (132106000073).

Author Contributions: In this work, the general concept has been developed by J.H., K.C., R.C. and S.W., while circuit boards for the data acquisition of the biosensing system have been constructed by H.L., J.L., M.J. and J.Z. Moreover, L.M., X.S. and X.H. have prepared the final draft and experiments have been performed by B.C. and H.L.

Conflicts of Interest: The authors declare no conflict of interest.

References

1. Abbas, A.; Linman, M.J.; Cheng, Q. New trends in instrumental design for surface plasmon resonance-based biosensors. *Biosens. Bioelectron.* **2011**, *26*, 1815–1824.
2. Perkins, E.A.; Squirrell, D.J. Development of instrumentation to allow the detection of microorganisms using light scattering in combination with surface plasmon resonance. *Biosens. Bioelectron.* **2000**, *14*, 853–859.
3. Piliarik, M.; Vaisocherová, H.; Homola, J. A new surface plasmon resonance sensor for high-throughput screening applications. *Biosens. Bioelectron.* **2005**, *20*, 2104–2110.
4. Gupta, G.; Sharma, P.K.; Sikarwar, B.; Merwyn, S.; Kaushik, S.; Boopathi, M.; Agarwal, G.S.; Singh, B. Surface plasmon resonance immunosensor for the detection of Salmonella typhi antibodies in buffer and patient serum. *Biosens. Bioelectron.* **2012**, *36*, 95–102.
5. Bolduc, O.R.; Live, L.S.; Masson, J.-F. High-resolution surface plasmon resonance sensors based on a dove prism. *Talanta* **2009**, *77*, 1680–1687.

6. Azzam, E.M.S.; Bashir, A.; Shekhah, O.; Alawady, A.R.E.; Birkner, A.; Grunwald, C.; Wöll, C. Fabrication of a surface plasmon resonance biosensor based on gold nanoparticles chemisorbed onto a 1,10-decanedithiol self-assembled monolayer. *Thin Solid Films* **2009**, *518*, 387–391.

7. Kajiura, M.; Nakanishi, T.; Iida, H.; Takada, H.; Osaka, T. Biosensing by optical waveguide spectroscopy based on localized surface plasmon resonance of gold nanoparticles used as a probe or as a label. *J. Colloid Interf. Sci.* **2009**, *335*, 140–145.

8. Bergström, G.; Mandenius, C.-F. Orientation and capturing of antibody affinity ligands: Applications to surface plasmon resonance biochips. *Sens. Actuators B* **2011**, *158*, 265–270.

9. Haughey, S.A.; Campbell, K.; Yakes, B.J.; Prezioso, S.M.; DeGrasse, S.L.; Kawatsu, K.; Elliott, C.T. Comparison of biosensor platforms for surface plasmon resonance based detection of paralytic shellfish toxins. *Talanta* **2011**, *85*, 519–526.

10. Mazumdar, S.D.; Barlen, B.; Kämpfer, P.; Keusgen, M. Surface plasmon resonance (SPR) as a rapid tool for serotyping of Salmonella. *Biosens. Bioelectron.* **2010**, *25*, 967–971.

11. Hu, J.D.; Hu, J.F.; Luo, F.K.; Li, W.; Jiang, G.; Li, Z.; Zhang, R. Design and validation of a low cost surface plasmon resonance bioanalyzer using microprocessors and a touch-screen monitor. *Biosens. Bioelectron.* **2009**, *24*, 1974–1978.

12. Endo, T.; Takizawa, H.; Imai, Y.; Yanagida, Y.; Hatsuzawa, T. Study of electrical field distribution of gold-capped nanoparticle for excitation of localized surface plasmon resonance. *Appl. Surf. Sci.* **2011**, *257*, 2560–2566.

13. François, A.; Boehm, J.; Oh, S.Y.; Kok, T.; Monro, T.M. Collection mode surface plasmon fibre sensors: A new biosensing platform. *Biosens. Bioelectron.* **2011**, *26*, 3154–3159.

14. Kim, Y.H.; Kim, J.P.; Han, S.J.; Sim, S.J. Aptamer biosensor for label-free detection of human immunoglobulin E based on surface plasmon resonance. *Sens. Actuators B* **2009**, *139*, 471–475.

15. Chinowsky, T.M.; Soelberg, S.D.; Baker, P.; Swanson, N.R.; Kauffman, P.; Mactutis, A.; Grow, M.S.; Atmar, R.; Yee, S.S.; Furlong, C.E. PorTable 24-analyte surface plasmon resonance instruments for rapid, versatile biodetection. *Biosens. Bioelectron.* **2007**, *22*, 2268–2275.

16. Mannelli, I.; Courtois, V.; Lecaruyer, P.; Roger, G.; Millot, M.C.; Goossens, M.; Canva, M. Surface plasmon resonance imaging (SPRI) system and real-time monitoring of DNA biochip for human genetic mutation diagnosis of DNA amplified samples. *Sens. Actuators B* **2006**, *119*, 583–591.

17. Myszka, D.G. Kinetic analysis of macromolecular interactions using surface plasmon resonance biosensors. *Curr. Opin. Biotechnol.* **1997**, *8*, 50–57.

18. Gnedenko, O.V.; Mezentsev, Y.V.; Molnar, A.A.; Lisitsa, A.V.; Ivanov, A.S.; Archakov, A.I. Highly sensitive detection of human cardiac myoglobin using a reverse sandwich immunoassay with a gold nanoparticle-enhanced surface plasmon resonance biosensor. *Anal. Chim. Acta* **2013**, *759*, 105–109.

19. Eum, N.-S.; Kim, D.-E.; Yeom, S.-H.; Kang, B.-H.; Kim, K.-J.; Park, C.-S.; Kang, S.-W. Variable wavelength surface plasmon resonance (SPR) in biosensing. *Biosystems* **2009**, *98*, 51–55.

84

20. Hu, J.; Li, W.; Wang, T.; Lin, Z.; Jiang, M.; Hu, F. Development of a label-free and innovative approach based on surface plasmon resonance biosensor for on-site detection of infectious bursal disease virus (IBDV). *Biosens. Bioelectron.* **2012**, *31*, 475–479.

21. Mitchell, J.S.; Wu, Y.; Cook, C.J.; Main, L. Sensitivity enhancement of surface plasmon resonance biosensing of small molecules. *Anal. Chem.* **2005**, *343*, 125–135.

22. Chabot, V.; Cuerrier, C.M.; Escher, E.; Aimez, V.; Grandbois, M.; Charette, P.G. Biosensing based on surface plasmon resonance and living cells. *Biosens. Bioelectron.* **2009**, *24*, 1667–1673.

23. Neff, H.; Zong, W.; Lima, A.M.N.; Borre, M.; Holzhüter, G. Optical properties and instrumental performance of thin gold films near the surface plasmon resonance. *Thin Solid Films* **2006**, *496*, 688–697.

24. Otsuki, S.; Ishikawa, M. Wavelength-scanning surface plasmon resonance imaging for label-free multiplexed protein microarray assay. *Biosens. Bioelectron.* **2010**, *26*, 202–206.

25. Gao, D.; Guan, C.; Wen, Y.; Zhong, X.; Yuan, L. Multi-hole fiber based surface plasmon resonance sensor operated at near-infrared wavelengths. *Opt. Commun.* **2014**, *313*, 94–98.

26. Davis, T.M.; Wilson, W.D. Determination of the refractive index increments of small molecules for correction of surface plasmon resonance data. *Anal. Chem.* **2000**, *284*, 348–353.

27. Kim, S.J.; Gobi, K.V.; Harada, R.; Shankaran, D.R.; Miura, N. Miniaturized portable surface plasmon resonance immunosensor applicable for on-site detection of low-molecular-weight analytes. *Sens. Actuators B* **2006**, *115*, 349–356.

28. Méjard, R.; Dostálek, J.; Huang, C.-J.; Griesser, H.; Thierry, B. Tuneable and robust long range surface plasmon resonance for biosensing applications. *Opt. Mater.* **2013**, *35*, 2507–2513.

29. Dong, W.; Pang, K.; Luo, Q.; Huang, Z.; Wang, X.; Tong, L. Improved polarization contrast method for surface plasmon resonance imaging sensors by inert background gold film extinction. *Opt. Commun.* **2015**, *346*, 1–9.

30. Jang, H.S.; Park, K.N.; Kang, C.D.; Kim, J.P.; Sim, S.J.; Lee, K.S. Optical fiber SPR biosensor with sandwich assay for the detection of prostate specific antigen. *Opt. Commun.* **2009**, *282*, 2827–2830.

31. Caide, X.; Sui, S.-F. Characterization of surface plasmon resonance biosensor. *Sens. Actuators B* **2000**, *66*, 174–177.

32. Aizawa, H.; Tozuka, M.; Kurosawa, S.; Kobayashi, K.; Reddy, S.M.; Higuchi, M. Surface plasmon resonance-based trace detection of small molecules by competitive and signal enhancement immunoreaction. *Anal. Chem.* **2007**, *591*, 191–194.

33. Chen, R.; Wang, M.; Wang, S.; Liang, H.; Hu, X.; Sun, X.; Zhu, J.; Ma, L.; Jiang, M.; Hu, J. A low cost surface plasmon resonance biosensor using a laser line generator. *Opt. Commun.* **2015**, *349*, 83–88.

34. Baccar, H.; Mejri, M.B.; Hafaiedh, I.; Ktari, T.; Aouni, M.; Abdelghani, A. Surface plasmon resonance immunosensor for bacteria detection. *Talanta* **2010**, *82*, 810–814.

35. Ashley, J.; Li, S.F.Y. An aptamer based surface plasmon resonance biosensor for the detection of bovine catalase in milk. *Biosens. Bioelectron.* **2013**, *48*, 126–131.

36. Bandyopadhyay, A.; Sarkar, K. Localized surface plasmon resonance-based DNA detection in solution using gold-decorated superparamagnetic Fe_3O_4 nanocomposite. *Anal. Chem.* **2014**, *465*, 156–163.

37. Bowen, J.; Meecham, J.; Hamlin, M.; Henderson, B.; Kim, M.; Mirjankar, N.; Lavine, B.K. Development of field-deployable instrumentation based on "antigen-antibody" reactions for detection of hemorrhagic disease in ruminants. *Microchem. J.* **2011**, *99*, 415–420.

Antigen-Antibody Affinity for Dry Eye Biomarkers by Label Free Biosensing. Comparison with the ELISA Technique

Maríafe Laguna, Miguel Holgado, Ana L. Hernandez, Beatriz Santamaría, Alvaro Lavín, Javier Soria, Tatiana Suarez, Carlota Bardina, Mónica Jara, Francisco J. Sanza and Rafael Casquel

Abstract: The specificity and affinity of antibody-antigen interactions is a fundamental way to achieve reliable biosensing responses. Different proteins involved with dry eye dysfunction: ANXA1, ANXA11, CST4, PRDX5, PLAA and S100A6; were validated as biomarkers. In this work several antibodies were tested for ANXA1, ANXA11 and PRDX5 to select the best candidates for each biomarker. The results were obtained by using Biophotonic Sensing Cells (BICELLs) as an efficient methodology for label-free biosensing and compared with the Enzyme-Linked Immuno Sorbent Assay (ELISA) technique.

Reprinted from *Sensors*. Cite as: Laguna, M.; Holgado, M.; Hernandez, A.L.; Santamaría, B.; Lavín, A.; Soria, J.; Suarez, T.; Bardina, C.; Jara, M.; Sanza, F.J.; Casquel, R. Antigen-Antibody Affinity for Dry Eye Biomarkers by Label Free Biosensing. Comparison with the ELISA Technique. *Sensors* **2015**, *15*, 19819–19829.

1. Introduction

As reported by Lemp *et al.* [1], dry eye disease is a multifactorial chronic disorder of the ocular surface that affects up to 100 million people worldwide. Diagnosis and management of dry eye has been a source of frustration to clinicians for a lack of correlation between signs and symptoms. Dry eye (DE) and meibomian gland dysfunction (MGD) are common inflammatory ocular surface diseases affecting tear film stability and ocular surface integrity. The pathophysiology of both conditions is complex and thought to represent the interaction of multiple mechanisms including tear film hyperosmolarity, instability, and subsequent activation of an inflammatory cascade, with release of inflammatory mediators into the tears, which in turn can damage the ocular surface epithelium.

Label-free optical biosensors have been demonstrated to be a good technology for *In-Vitro* Diagnostics (IVD) due to advantages *versus* labeled techniques [2,3]. The short turnaround and cost-effectiveness advantages are very important factors for final users and health professionals as a whole. Mainly, three important factors are connected with the Limit of Detection (LoD) of optical label-free biosensing: the transducer sensitivity, resolution of the optical reader and the performance of the immunoassay. The latter one, the antigen-antibody interaction, plays an important role to achieve a competitive LoD. In this sense, the study of specificity and affinity of antibody-antigen interactions is fundamental for understanding the biological activity of these proteins, as well as to develop suitable biosensors.

As it is well explained [4,5], a highly specific bimolecular association is achieved by the interaction between an antibody with its corresponding antigen, which involves various non-covalent interactions between the antigen epitope and the variable region of the antibody molecule. These interactions (ionic bonds, hydrogen bonds, hydrophobic interactions and van der Walls interactions) are needed for a strong antigen-antibody binding requiring a high degree of complementarity between antigen (Ag) and antibody (Ab).

Affinity is the strength of binding of a single molecule to its corresponding ligand. Typically it is determined by the equilibrium dissociation constant (K_D), which is used to evaluate biomolecular interactions. The measurement of the reaction rate constants can be used to define an equilibrium or affinity constant ($1/K_D$). Thus, the smaller the K_D value, the greater the affinity of an antibody with its target. Antibodies with high affinity have an association constant $K_a > 10^7$ M^{-1} [6,7].

Biomarkers are frequently used in clinical trials of therapeutics for the assessment of disease states and also for evaluating diagnostic devices. In previous works, several biomarkers where validated for dry eye disease: S100A6, CST4, MMP9, PRDX5, ANXA1, ANXA11, PLAA [8].

In previous articles, our research group has also proven an efficient methodology for label-free biosensing by using Biophotonic Sensing Cells (BICELLs) [9,10], and particularly for dry eye diseases [11]. According to this, in this article we study the affinity of several antibodies for biomarkers: ANXA1, ANXA11, PRDX5 and S100A6 using BICELLs based on SU8 resist Fabry-Perot interferometers with an optical read-out of the biosensor based on the interferometry.

The label-free optical technique based on BICELLs is a well-reported optical technique where basically changes in the refractive index are produced by the recognition or accumulation events of biomolecules onto the sensing surface [9]. This BICELLs method is a label-free, which means that it is not necessary label-molecules for the detection. However, in the classical Enzyme-Linked Immuno Sorbent Assay (ELISA) protocols a labeled-molecule for subsequent detection is needed.

2. Experimental Section

2.1. Production of Mouse mAbs

The mAbs were obtained from female Balb/c mice immunized by intraperitoneal injections with the recombinant proteins ANXA1, ANXA11 and PRDX5, separately. The fusion was performed using a Clona Cell-HY kit following the manufacturer's instructions (Stemcell Technologies, Vancouvert, BC, Canada). Briefly, micesplenocytes were fused with immortal NSO-1 cells (kindly donated by Margaret Goodall, University of Birmingham, Birmingham, UK) with the addition of polyethylene glycol (Clona Cell-HY kit). The resulting mix was grown in selective agar (ClonaCell-HY kit) on 96-well plates.

Screening of positive hybridoma cell culture supernatant was tested by indirect ELISA. Desired clones were expanded, cultured on a large scale and cryopreserved. The three best hybridomas of each fusion were selected (Table 1) based on its productivity, ELISA signal and growth rate for further studies.

mAbisotypes were determined with the mouse mAbisotyping kit (Sigma-Aldrich, Madrid, Spain), and were purified by Protein G (GE Healthcare, Buckinghamshire, UK) affinity column chromatography. Their purity was confirmed by SDS/PAGE. All mAbs were produced and purified by AntibodyBcn (Barcelona, Spain).

Table 1. Antibodies selected from each fusion.

Protein	Antibody Selected
ANXA 1	P4D1
	P6D7
	P10B12
ANXA11	P1B11
	P3F9
	P4D9
PRDX5	P3G1
	P5H6
	P9F4

2.2. Affinity ELISA Assay

In order to establish which mAb shown a greater affinity to its own antigen, calibrating curves were carried out by indirect ELISA assays as follows. Ninety-six-well ELISA plates (Santa Cruz Biotech, Dallas, TX, USA) were coated for 4 h at 37 °C with 100 μL per well of each protein in serial dilutions (1:2) from 200 ng/mL to 3.125 ng/mL in 0.2 M carbonate buffer (pH 9.6). Washing was done using 0.05% Phosphate Buffer Saline (PBS)-Tween 20 (PBS-T). Wells were blocked with 2.5% non-fat milk-PBST overnight at 4 °C. Afterwards plates are incubated with 100 μL purified mAbs at 5 μg/mL for 1 h at 37 °C. Ab binding was detected with HRP-conjugated anti-mouse IgG (HRP stands for Horseradish Peroxidase; 1:500 in PBS-T; Santa Cruz Biotech), followed by color development with tetramethylbenzidine ELISA substrate (TMB; Thermo Fisher Scientific, Uppsala, Sweden). The reaction was stopped with 1 M HCl and read at 450 nm by a Multiscan FC microplate reader (Thermo Fisher Scientific).

2.3. Biosensor

For this experimental work we used, as photonic transducer, a Biophotonic Sensing Cell (BICELL) based on Fabry-Perot interferometers of SU8 polymeric resist that exhibits a sensitive optical label-free biosensing capability. The Fabry-Perot interferometer is the biotransducer of the biosensor itself. Bicells are based on different type of interferometers and are normally square sensing areas where the recognition events take place. For this particular case, the interferometer employed is a single SU8 layer Fabry-Perot interferometer where part of the light is transmitted through the SU8 reaching the substrate. As a result the interference is produced by the mixed beams coming from the SU8 (and its biomolecules) and the substrate. The large number of interfering beams produces an interferometry profile with a high resolution suitable for biosensing.

We employed SU8 2000.5 (Microchem Corp., Newton, MA, USA) diluted in cyclopentanone [12] for the fabrication of BICELLs. The SU8 resist was deposited by spinning at 3000 rpm for 3 min, then the film was soft-baked at 70 °C for 1 min. An exposure to UV light process was then carried out, followed by a post-bake step at 70 °C for 5 min in order to give a stable thin film. The SU8 surface of the BICELLs was treated with sulfuric acid (95% for 10 s) in order to have a hydrophilic sensing surface. As a result of this treatment, the SU8 epoxy groups are opened and suitable to immobilize covalently the protein [13].

By monitoring the changes in the interferometric profile of theoptical mode response, the immobilization of protein and the recognition of several antibodiescan be properly monitored. Therefore, it is possible to detect the response of the antibody for each biomarker.

2.4. Optical Characterization of the Biosensor and Sensing Principle

The optical readout of the biosensor was accomplished by a Fourier transform visible-infrared (FT-VIS-IR) spectrometer (Vertex 70 adapted to the visible range, Bruker, Madrid, Spain) after each incubation/washing step. We followed the well-described procedure very recently reported in the literature [9] (see in Figure 1a–c).

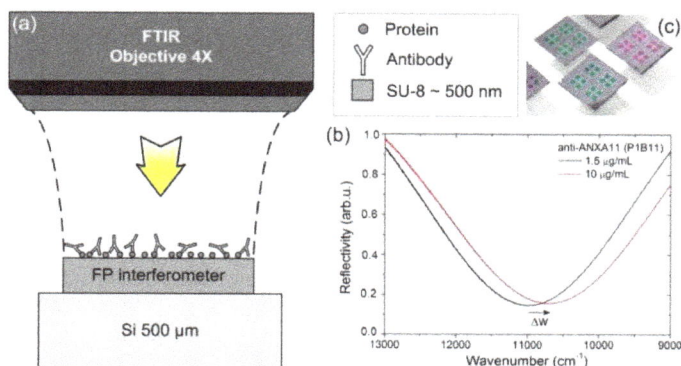

Figure 1. (a) Optical setup for measurements and biochemical diagram of the immunoassay; (b) optical response for the BICELLs; (c) Bicells used in the immunoassay.

2.5. Immunoassay Procedure

The indirect immunoassay Protein (ANXA1, ANXA11 and PRDX5)/antibody was carried out by a covalent binding of the protein onto the BICELLs SU8 sensing surface until saturation for testing the best clone obtained for AntibodyBcn (Barcelona, Spain). The covalent bond occurs between epoxy ring of SU8 and amine groups of proposed proteins. The incubation of proteins was made until saturation with a volume of 60 μL, with a concentration of 50 μg·mL^{-1} in phosphate buffered saline (pH 7.4,), and at temperature of 37 °C during 20 min. Then, the surfaces were rinsed with deionize water (DI-H$_2$O) and blown with dry and clean dust-less air under clean environment.

Avoid nonspecific adsorption is a very important step. In fact, the blocking step avoids the unspecific bounding, especially important for direct immunoassay, where the antibody is firstly immobilized onto the sensing surface. However, for this article, we did not consider using a blocking step because we immobilized the biomarker(indirect immunoassay) until saturation, supposing that the sensing surface is completely filled with the protein (there are a biofilm of protein according with our previous simulations).

Then, we proceeded to recognize the corresponding antibody. The recognition curve of antibody with concentrations 0.2, 0.5, 1.5, 2.5, 5, 10, 25, 5, 10, 25, 50 and 100 μg·mL^{-1} in PBS-pH 7.4 was observed at 37 °C for 20 min for each incubation step. Thus, for each antibody concentration the corresponding BICELLs were washed with PBS-T and water and blown with dry and clean dust-less air.

3. Results and Discussion

3.1. Results Obtained by ELISA Technique: Affinity Analysis by ELISA

Selected monoclonal antibodies were individually characterized to determine which of them showed the highest affinities which meant strong binding ability to their antigen and would lead to its strong applied value in areas such as detection and diagnosis. Thus, for ANXA 1 (Figure 2a), the antibody P10B12 did not show a significant signal even at high ligand concentrations. The other two antibodies shown a slightly improvement, being antibody P6D7 a little better than the antibody P4D1, with dissociation constants K_D of 2.40 μM, and 27.01 μM, respectively. Both antibodies give signals too far from the saturation range, however both antibodies could be used for ANXA1 detection.

In the case of monoclonal antibodies against protein ANXA11 (Figure 2b) all of them showed apparently good signals; both P3F9 and P1B11 are close to the saturating point at the highest ligand concentration employed in the assay. Although P3F9 demonstrated the best ability to bind to antigen ANXA11, P1B11 and P4D9 could be also used for an effective detection of the protein. The dissociation constant (K_D) of P3F9, P4D9 and P1B11 were 19 nM, 4.87 μM and 1.56 μM, respectively.

Finally in the case of antibodies the intensity shown by the three selected antibodies against PRDX5 (Figure 2c) reveals a high affinity of all of them. Antibody P9F4 has the higher affinity to PRDX5 with a K_D of 17.66 nM. Both P3G1 and P5H6 antibodies have a similar affinity rate with a K_D of 22.05 nM and 27.01 nM, respectively.

3.2. Results Obtained by Optical Label-Free Technique

In order to analyze the response of the antibody for each biomarker, we evaluated the spectral response for different concentrations of antibody. Figure 3 shows the measured interference dip wavenumber displacement of Fabry-Perot interferometer for increasing concentrations of the different antibodies. In the analyte-receptor recognition reaction, the dissociation constant is expressed as $K_D = [A] \cdot [R]/[AR]$, where [A] is the free analyte concentration, [R] is the free receptor concentration and [AR] is the analyte-receptor complex concentration. At the equilibrium, $K_D = k_d/k_a$, k_d and k_a are the kinetic constants for the dissociation and association process, respectively. Thus, K_D can be considered as the reciprocal of the analyte affinity towards the receptor. In our experiment the receptor concentration is assumed to be $[R] = [R]_{total} - [AR]$ and when 50% of the binding sites are occupied ($[AR] = 0.5 \cdot [R]_{total}$), the dissociation constant is the free analyte concentration $K_D = [A]$. Therefore, the K_D value is the antibody concentration causing a response in the transduction equal to 50% of the total transduction change after saturation. In Figure 3a (for ANXA1) two clones were studied (P4D1 and P6D7). The signal for

P4D1 clone is much lower than P6D7 clone. Both clones gave an affinity constant values very low (P6D7-K_D = 1.6 × 10^{-4} M and P4D1 = 8.86 × 10^{-5} M), resulting in a poor affinity for the protein ANXA1 because antibodies with high affinity must have K_D < 10^{-7} M. For these reasons, both antibodies are not considered very good for recognizing the ANXA1 biomarker.

For Anxa11 (Figure 3b) he three antibodies offered a good dynamic range with dissociation constant values lower than 10^{-7} M. The values obtained for P3F9, P4D9 and P1B11 are 20 nM, 15 nM and 33.3 nM, respectively. Figure 3b shows that all antibodies reach the point of saturation below 10 µg· mL^{-1} and the dissociation constants values obtained show the high affinity of the antibodies to its corresponding antigen.

Finally for PRDX5 (Figure 3c), three selected antibodies were studied, showing a high affinity towards PRDX5. The dissociation constants values obtained for P9F4, P5H6 and P3G1 are 7.3 nM, 23.3 nM and 26.6 nM, respectively. These values are in agreement with values obtained by the ELISA technique.

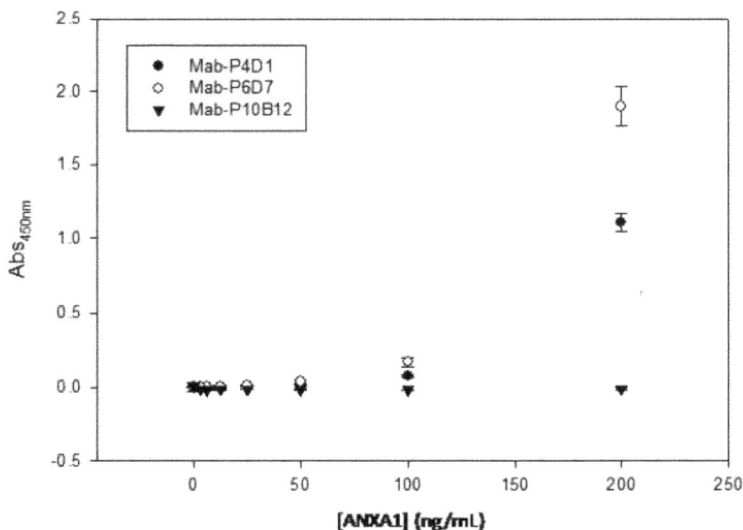

(a) ANXA1

Figure 2. *Cont.*

93

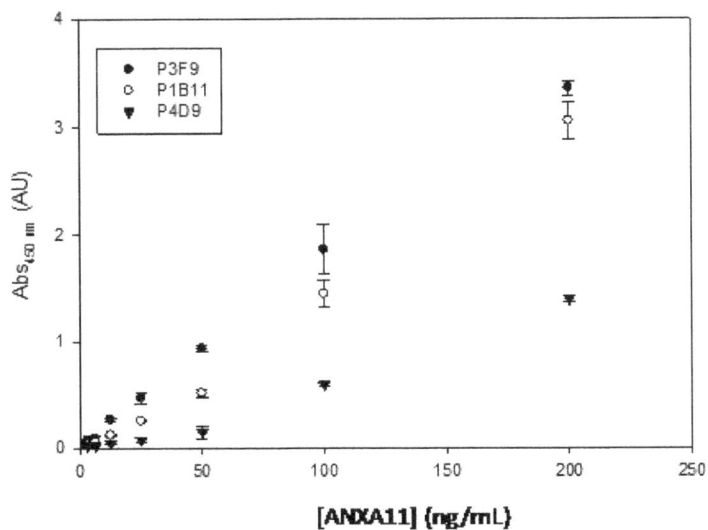

(b) ANXA11

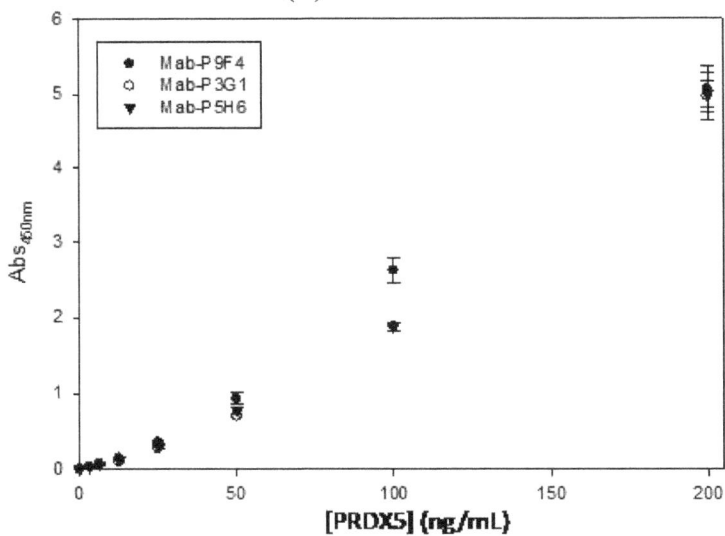

(c) PRDX5

Figure 2. Calibration curves for selected monoclonal antibodies. The absorbance measurements are plotted against the protein concentration ranging from 3.125 ng·mL^{-1} to 200 ng·mL^{-1}: (**a**) ANXA1; (**b**) ANXA11; (**c**) PRDX5.

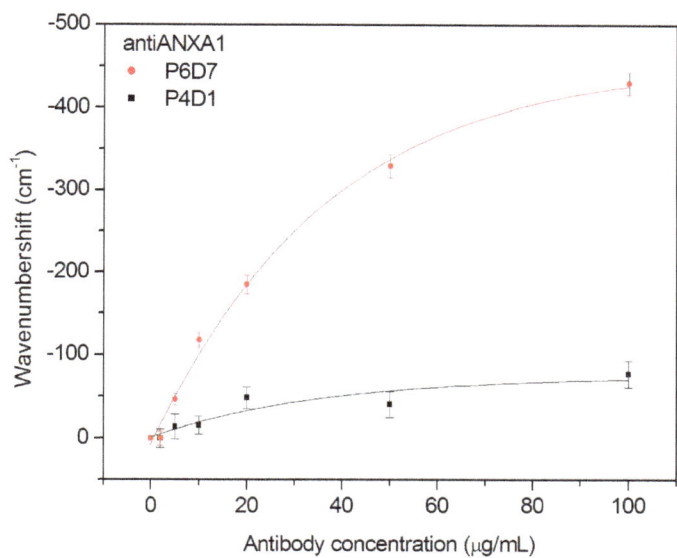

(a) ANXA1

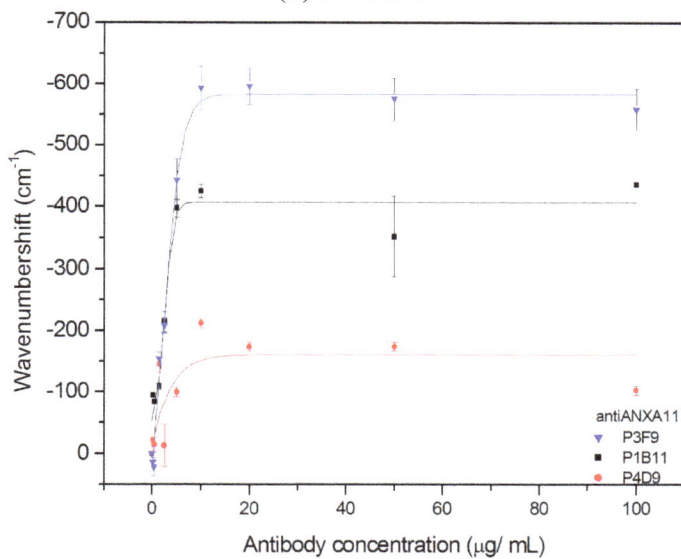

(b) ANXA11

Figure 3. Cont.

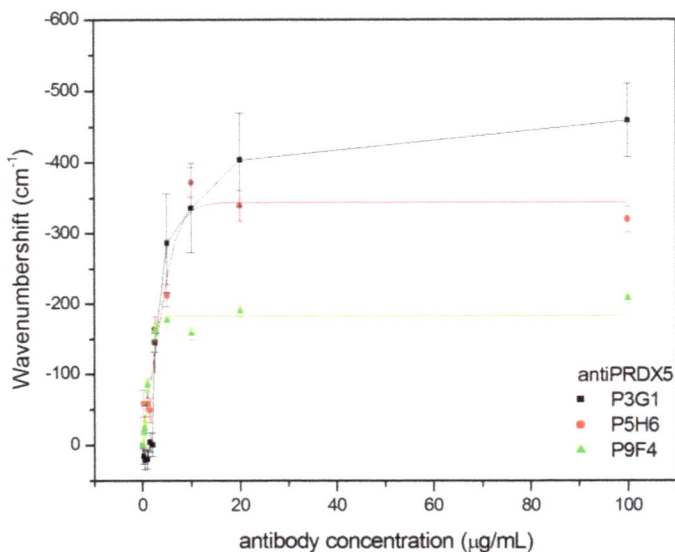

(c) PRDX5

Figure 3. Dip shift against antibody concentration for (**a**) ANXA1; (**b**) ANXA11; (**c**) PRDX5.

3.3. Comparison of the ELISA Technique versus the Optical Label-Free Technique

Dissociation constant values for eight antibody-antigen systems were compared with ELISA and the Optical Label-Free technique. This analysis, shown in Figure 4, indicated that five antibodies have K_D values in the same order of magnitude with both techniques. However, three antibodies show dissociation constant values that differ by two orders of magnitude.

The differences between both techniques can be justified as follows: the Enzyme-Linked Immuno Sorbent Assay (ELISA) technique is a method where affinity constants is determined in dilution and, therefore, a real immunoreaction constant is determined. However, by employing the optical interferometric technique based on BICELLs, the reaction constant is calculated in the solid-phase, leading an apparent constant in a heterogeneous biosensing assay. Moreover, the optimization of immunoassay (e.g., pH of buffers, incubation times, and temperature, among others) may have significant implications and influence the antigen-antibody interaction. For this reason, the quantitative estimation of the affinity constant with our optical interferometric technique is an essential piece of information when setting up a heterogeneous biosensing assay.

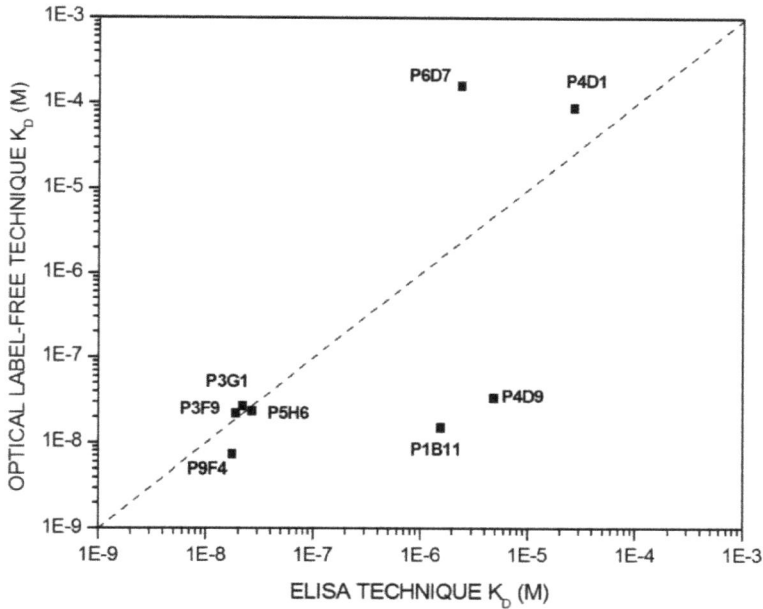

Figure 4. Equilibrium dissociation constant (K_D) measurements obtained using the Elisa technique and the Optical Label-free technique.

4. Conclusions

The affinity antigen-antibody for several biomarkers associated with dry eye disease was studied using an optical label-free interferometric technique. For this study biophotonic sensing cells (BICELLs) based on SU8 photoresist have been used.

Antibodies for three biomarkers: ANXA1, ANXA11 and PRDX5 were produced. The affinity of the antibodies was tested by the ELISA technique calculating their dissociation constant (K_D) and therefore the affinity to their corresponding antigens.

An indirect immunoassay until antigen saturation on the sensing surface took place by an optical label-free technique was performed. Then, a recognition curve for each antibody was plotted. From this curve, an apparent dissociation constant (K_D) was calculated and compared with the ELISA result.

In general terms, antibodies with $K_D < 10^{-7}$ M have high affinity. Therefore, for the ANXA1 biomarker, two antibodies were studied by using ELISA and the optical label-free technique. As a result, both antibodies exhibited poor affinity. However, for the ANXA11 biomarker we observed a good affinity reaction: the best antibody is P3F9 for both techniques. Finally, for the PRDX5 biomarker the three antibodies also had a good affinity by both techniques.

As a main conclusion, the comparative analysis of K_D indicates a reasonable correlation between both techniques in some antigen-antibody pairs. However, in other pairs there are significant differences. We consider that the main different values of K_D between both IVD techniques are more related with the different immunoassays protocols when using ELISA in solution in comparison with the BICELLs based optical interferometric technique in heterogeneous medium. As explained above, parameters such as buffer, sample volume, incubation time, blocking steps and washing can impact the determination of the K_D. Finally, even with the different K_D values observed, the proposed interferometric optical label-free technique seems to be suitable to study antigen-antibody affinity.

Acknowledgments: This work has been supported under the Project INNBIOD (REF: IPT-2011-1429-010000) under the Spanish Ministry of Economy and Competitiveness (INNPACT program), the Project PLATON (REF: TEC2012-31145) under the Spanish Ministry of Economy and competitiveness (Plan Nacional 2012 Investigación fundamental no orientada) and ENVIGUARD EUROPEAN PROJECT FP7-OCEAN-2013. We acknowledge specially Bioftalmik and Antibody Bcn for theirtechnical and scientific support, as well as for the bio-chemical reagents to carry out the experiments described in this article.

Author Contributions: Maríafe Laguna and Miguel Holgado designed the experiments and performed the optical experiments to prove the concept. Francisco J. Sanza designed the photonic transducer. Ana López and Beatriz Santamaría carried out the optical biosensing measurements. Alvaro Lavín and Rafael Casquel wrote the theoretical simulation codes for this experiments. Carlota Bardina and Mónica Jara obtained the mouse mAbs. Javier Soria and Tatiana Suarez carried out the ELISA assays.

Conflicts of Interest: The authors declare no conflict of interest.

References

1. Lemp, M.A.; Sullivan, B.D.; Crews, L.A. Biomarkers in Dry Eye Disease. *Eur. Ophthalmic Rev.* **2012**, *6*, 157–163.
2. Fan, X.; White, I.M.; Shopova, S.I.; Zhu, H.; Suter, J.D.; Sun, Y. Sensitive optical biosensors for unlabeled targets: A review. *Anal. Chim. Acta* **2008**, *620*, 8–26.
3. Esteves, M.C.; Alvarez, M.; Lechuga, L.M. Integrated optical devices for lab-on-a chip biosensing applications. *Laser Photonics Rev.* **2012**, *6*, 463–487.
4. Laudry, J.P.; Fei, Y.; Rhu, X. Simultaneous measurement of 10000 protein-ligand affinity constants using microarray-based kinetic constant assays. *Assay Drug Dev. Technol.* **2012**, *10*, 250–259.
5. Rispens, T.; Te, V.H.; Hemker, P.; Speijer, H.; Hermens, W.; Aarden, L. Label-free assessment of high-affinity Ab-Ag binding constanst: Comparison of bioassay, SPR and PEIA-ellipsometry. *J. Immunol. Methods* **2011**, *365*, 50–57.
6. Goldsby, R.A.; Kindt, T.K.; Osborne, B.A.; Kuby, J. *Immunology*, 5th ed.; W.H. Freeman and Company: New York, NY, USA, 2003.
7. K_D value: A Quantitative Measurement of Antibody Affinity. Available online: www.abcam.com (accessed on 6 August 2015).

8. Soria, J.; Duran, J.A.; Etxebarria, J.; Merayo, J.; Gonzalez, N.; Reigada, R.; García, I.; Acera, A.; Suarez, T. Tear proteome and protein network analyses reveal a novel pentamarker panel for tearfilm characterization in dry eye and meibomian gland dysfunction. *J. Proteomics* **2013**, *78*, 94–112.

9. Sanza, F.J.; Holgado, M.; Ortega, F.J.; Casquel, R.; López-Romero, D.; Laguna, M.F.; Bañuls, M.J.; Barrios, C.A.; Puchades, R.; Maquieira, A. Bio-Photonic Sensing Cells over transparent substrates for anti-gestrinone antibodies biosensing. *Biosens. Bioelectron.* **2011**, *26*, 4842–4847.

10. Holgado, M.; Barrios, C.A.; Sanza, F.J.; Casquel, R.; Laguna, M.F.; Bañuls, M.J.; López-Romero, D.; Puchades, R.; Maquieira, A. Label-free biosensing by means of periodic lattices of high aspect ratio SU-8 nanopillars. *Biosens. Bioelectron.* **2010**, *25*, 2553–2558.

11. Laguna, M.F.; Sanza, F.J.; Soria, J.; Jara, M.; Lavín, A.; Casquel, R.; Lopez, A.; Suarez, T.; Holgado, M. Label-free biosensing by means of BICELLs for dry eye. *Sens. Actuators B Chem.* **2014**, *203*, 209–212.

12. Del Campo, A.; Greiner, C. SU-8: A photoresist for high-aspect-ratio and 3D submicron lithography. *J. Micromech. Microeng.* **2007**, *17*, 81–95.

13. Ortega, F.J.; Bañuls, M.-J.; Sanza, F.J.; Laguna, M.F.; Holgado, M.; Casquel, R.; Barrios, C.A.; López-Romero, D.; Maquieira, A.; Puchades, R. Development of a versatile biotinylated material based on SU-8. *J. Mater. Chem. B* **2013**, *1*, 2750–2756.

PEG Functionalization of Whispering Gallery Mode Optical Microresonator Biosensors to Minimize Non-Specific Adsorption during Targeted, Label-Free Sensing

Fanyongjing Wang, Mark Anderson, Matthew T. Bernards and Heather K. Hunt

Abstract: Whispering Gallery Mode (WGM) optical microresonator biosensors are a powerful tool for targeted detection of analytes at extremely low concentrations. However, in complex environments, non-specific adsorption can significantly reduce their signal to noise ratio, limiting their accuracy. To overcome this, poly(ethylene glycol) (PEG) can be employed in conjunction with appropriate recognition elements to create a nonfouling surface capable of detecting targeted analytes. This paper investigates a general route for the addition of nonfouling elements to WGM optical biosensors to reduce non-specific adsorption, while also retaining high sensitivity. We use the avidin-biotin analyte-recognition element system, in conjunction with PEG nonfouling elements, as a proof-of-concept, and explore the extent of non-specific adsorption of lysozyme and fibrinogen at multiple concentrations, as well as the ability to detect avidin in a concentration-dependent fashion. Ellipsometry, contact angle measurement, fluorescence microscopy, and optical resonator characterization methods were used to study non-specific adsorption, the quality of the functionalized surface, and the biosensor's performance. Using a recognition element ratio to nonfouling element ratio of 1:1, we showed that non-specific adsorption could be significantly reduced over the controls, and that high sensitivity could be maintained. Due to the frequent use of biotin-avidin-biotin sandwich complexes in functionalizing sensor surfaces with biotin-labeled recognition elements, this chemistry could provide a common basis for creating a non-fouling surface capable of targeted detection. This should improve the ability of WGM optical biosensors to operate in complex environments, extending their application towards real-world detection.

Reprinted from *Sensors*. Cite as: Wang, F.; Anderson, M.; Bernards, M.T.; Hunt, H.K. PEG Functionalization of Whispering Gallery Mode Optical Microresonator Biosensors to Minimize Non-Specific Adsorption during Targeted, Label-Free Sensing. *Sensors* **2015**, *15*, 18040–18060.

1. Introduction

Biosensors combine biological components with traditional physicochemical detection systems that operate via optical, electrical, or mechanical signal transduction mechanisms, offering advantages in the specific and timely detection of

100

biomolecular species. Optical biosensors can be classified into two types of sensors: labeled optical biosensors, such as the fluorescence-based family of biosensors, and label-free optical biosensors, such as the refractometric family of biosensors. Labeled biosensors rely on the detection of the label, rather than the biomolecular species of interest, while label-free biosensors theoretically have a high enough signal to noise ratio (SNR) that they are capable of directly detecting the biomolecular species of interest [1]. Of these, labeled biosensors, like fluorescence-based biosensors, are the most widely-used [2], but they have a number of disadvantages including the requirements of labeling, the cost of the peripheral equipment needed to perform the detection, and the possible difficulties in conjugation and quantification due to the presence of the fluorescent label [3,4]. Label-free optical biosensors, on the other hand, may not only overcome these limitations but may also have the potential to deliver higher quality and resolution detection, with more information content and fewer false negatives, as compared to labeled biosensors [5]. Typically, these platforms use a high-sensitivity signal transducer to convert a stimulus-induced response into a quantifiable signal, without relying on dyes, enzymes, or radiolabels [5,6]. The most attractive feature of this type of biosensor is that the detection could be performed on-site and in real-time, without the need for additional peripheral equipment [5]. Due to these advantages, label-free optical biosensors have been used widely in many fields, such as medical diagnostics, drug screenings, food safety, environmental protection, biotechnology assays, and biohazard security screenings [7,8]. Moreover, they are playing an essential role in ultra-low detection studies, as well as studies designed to understand the interactions between and among biomolecular species [9,10].

One example of a refractometric optical device that is capable of performing label-free biosensing is the Whispering Gallery Mode (WGM) optical microresonator [11–13]. The Whispering Gallery Mode is a morphology-dependent resonance that has unique properties including low cavity loss [1]. To create a label-free biosensor from this device, the WGM optical microresonator must be excited via light from an external source. A necessary precondition of the occurrence of the WGM is that the dielectric optical microresonator must have a higher refractive index than the surrounding media, so that light can be spatially confined in the resonator by total internal reflection (TIR) and propagate along the microresonator's periphery at specific resonant frequencies. The repeated reflection around the boundary of the microresonator results in the creation of an evanescent field in the surrounding environment that decreases exponentially with the distance away from the interface [14]. This field allows the microresonator to interact with biomolecules in the surrounding environment. Molecules adsorbing or binding onto the microresonator will cause a slight deviation in the effective refractive index of the circulating optical field, resulting in a detectable shift in the resonant frequency of the optical field contained by the device (Figure 1). Note that WGM positions are very sensitive to any modification of the refractive index of either the resonator or the surrounding media [15,16]. It is these interactions that give the device its sensing capabilities.

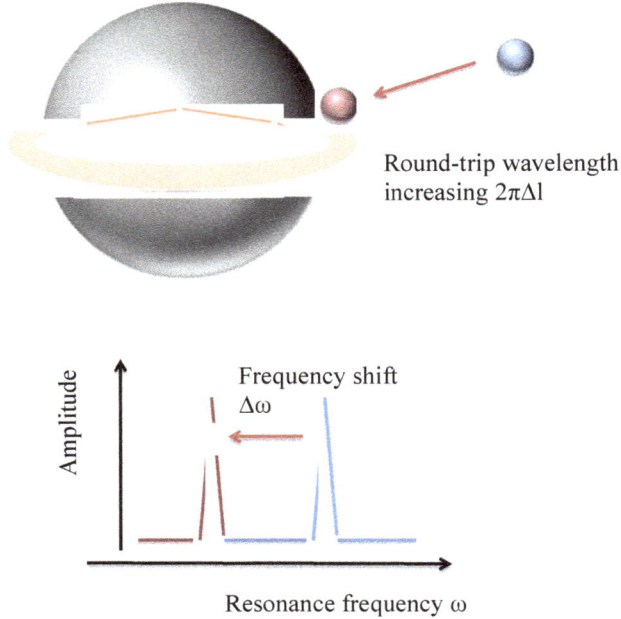

Figure 1. Illustration of WGM resonator, based on [17] (adapted with permission). Light (orange) enters a WGM resonator, where it experiences total internal reflection (TIR) and generates an evanescent field. When an analyte binds or adsorbs onto the surface of the microsphere, it changes the effective refractive index of the circulating optical field resonator, and it pulls part of the evanescent field to the outside of the resonator (blue for pre-binding, red for post-binding). The expansion of the optical field's boundary causes the round-trip wavelength of light to increase about $2\pi\Delta l$. The increase in the light wavelength results in a frequency shift in the transmission spectrum. The evanescent field is an optical field extending to the surrounding environment and decreasing exponentially with the distance away from the resonator's interface.

WGM optical microresonators can be fabricated in different geometries, such as microrings, microdisks, microtoroids, microspheres, microcylinders, or even asymmetric optical cavities [18–21]. However, most biosensors based on the WGM optical microcavities have been fabricated in the microsphere geometry from single-mode, silica optical fiber, typically resulting in an average diameter of 200 μm [22]. The most attractive property of the WGM optical microresonator is that both intrinsic and coupling loss can be extremely low. The total loss in the device is the primary factor in determining the maximum sensitivity of the resulting biosensor: lower intrinsic loss results in a longer photon lifetime in the microresonator. It, in turn, enhances the resulting interaction between the circulating photons and biomolecules on the microresonator's surface [14,23]. Due to their high sensitivity, compact structure, and simple fabrication, WGM optical microresonators have been employed as an efficient platform for molecular detection, single-atom detection, and temperature measurement [24].

However, one of the primary challenges of utilizing label-free biosensors, like WGM optical microresonator-based biosensors, is the lack of a recognition element to provide specific or selective detection, in addition to sensitive detection. In order to perform either selective or specific detection, these biosensors must be combined with some kind of recognition element that allows them to selectively or specifically target a biomolecular species of interest. This recognition element could be an antibody, a protein, an enzyme, or even a functional organic group [25]. Once this task has been accomplished, the biosensor must then be capable of specific and sensitive detection in complex environments, where non-specific adsorption may occur, reducing the SNR and the overall platform performance. One of the primary challenges for label-free biosensors, and indeed, for biosensors in general, is the non-specific adsorption of unwanted biomolecules when the biosensors are used in complex environments, such as water samples, blood, and serum [1].

Although there are many techniques that may be used to reduce non-specific adsorption on biosensors, the most popular is the use of "blocker elements," or "nonfouling elements", as a component of the surface chemistry applied to the device to make it selective or specific. This can be done via physical adsorption, self-assembled monolayers, or covalent binding. The use of different deposition or functionalization techniques depends on the outcome desired; in many cases, the stability of a covalently-bound nonfouling element, particularly in terms of time and temperature, makes them more attractive candidates than physically adsorbed layers, despite their potential additional difficulty. Fortunately, there is also a wide variety of nonfouling elements available, allowing researchers to find a nonfouling chemistry that works best for a specific application. These include nonfouling elements, such as bovine serum albumin (BSA), lipids, non-ionic detergents like Tween 20, and of course, polymeric materials, like polyethylene glycol (PEG), as well as combinations of these elements [26–29]. PEG is one of the most-studied and general polymeric materials for nonfouling coatings, especially in the pharmaceutical, cosmetic, and biomedical fields [30]. The interest in this polymer is driven by its unique physical, chemical, and biological properties in conjunction with its behavior towards proteins and other biologically-active molecules. This includes its excellent solubility in both aqueous and organic media, and non- immunogenicity, antigenicity, or low toxicity towards living cells [31,32]. One of the most widespread applications for PEG is to resist non-specific adsorption via its strong hydration layer and steric stabilization effect [31–33]. This concept has been used, for instance, in biomaterials to create nonfouling surfaces, as the attachment alters the electric nature of the surface exposed to the surrounding fluids [33]. When used in conjunction with biosensor platforms, the modification by PEG is used to obtain a significant reduction in the non-specific interaction of biological molecules with the biosensor's surface, because PEG is highly hydrophilic and has appreciable chain flexibility [34]. Non-specific adsorption decreases the SNR by increasing the background noise, and thus degenerates the sensing ability, even in high sensitivity biosensors. As mentioned above, a biosensor typically consists of a high-specificity recognition element and a high-sensitivity transducer. The PEG coatings should prevent non-specific biomolecules from binding to the surface, while allowing the recognition elements to bind with the targeted analytes, thus improving the overall performance of the biosensor.

In our previous work, we explored the use of PEG coatings in combination with WGM optical microresonator biosensors to minimize non-specific adsorption of fibrinogen and lysozyme during non-targeted detection [35]. In that work, PEG coatings of varying molecular weight were attached to the biosensor surface and were proven to have the capability to reduce non-specific adsorption. It was found that the short-chain PEG surfaces performed better in minimizing non-specific adsorption compared with long-chain PEG surfaces [35]. Here, we extend this work to targeted sensing using the biotin-avidin recognition element-target system as a proof of concept. The reasoning for the use of this system is that the biotin-avidin-biotin complex is frequently used as an intermediate sandwich complex when functionalizing surfaces of sensors; by first grafting biotin to the surface, then associating it with avidin, numerous biotin-labeled recognition elements can then be bound to the surface using the high affinity of avidin for biotin. The chemistry presented here, then, could be used as a general approach to reducing non-specific adsorption for targeted sensing using many different recognition elements. We evaluate the capability of different PEG-biotin:PEG ratios (1:1, 1:2, 1:3) in preventing non-specific adsorption, hypothesizing that the amount of exposed (PEG only) nonfouling elements would significantly impact the amount of non-specific adsorption. For each ratio chosen, the amount of the biotin recognition element (PEG-biotin) in solution was held constant while the amount of PEG nonfouling elements was increased. Fibrinogen and lysozyme were used to test for non-specific absorption to the PEG-biotin:PEG coated microresonators. The results show that these two proteins interacted minimally with the coated microresonator. Avidin was then used to test for a specific interaction. The results demonstrate that the PEG-biotin:PEG coated microresonator can effectively recognize avidin in a concentration dependent manner.

2. Experimental Section

PEG plays an important role in resisting non-specific protein binding by creating a nonfouling surface on the microsphere. In addition, the use of biotin as the recognition element allows the creation of sandwich complexes with avidin and recognition elements, such as antibodies, proteins, receptors, *etc.*, labeled with biotin. Therefore, the combination of PEG and PEG-biotin as a coating on the microspheres should improve their specificity to the target molecules by rejecting detrimental protein molecules, and thus reducing the occurrence of false positives (Figure 2). The different ratios of PEG-biotin to PEG indicate the different densities of the nonfouling elements on the surface. We investigated if the presence of differing densities of nonfouling elements could make a significant impact on the capability of resisting non-specific adsorption.

2.1. Synthesis and Characterization of Functionalized (100) Silica on Silicon Wafers

To investigate the resistance of the functionalized surfaces towards non-specific adsorption, we used both (100) Si wafers with a 2 μm thermal oxide layer of SiO_2 (University Wafers) as a control surface, as many typical surface characterization techniques have difficulty evaluating 3D curved surfaces accurately, as well as the silica microsphere optical microresonators (Section 2.2). Here, the thickness of the coating on the wafers was measured both before and after adsorption using ellipsometry. Additionally, optical profilometry and contact angle measurement were also used to investigate the surface quality and hydrophobicity characteristics. By comparing the thickness change due to the adsorption, the ratio of PEG-Biotin to PEG that demonstrated the best nonfouling characteristics could be selected and then applied to the three-dimensional optical microresonators.

To do this, silica-on-silicon (100) wafers (University Wafer) with a 2 μm silica (thermal oxide) grown on the surface were cut into rectangular pieces of 2 cm × 0.8 cm. Five different sets of chemistries were applied to these wafers: PEG and PEG-biotin were deposited on the wafer pieces with each of the three ratios (1:1, 1:2, and 1:3, PEG-biotin:PEG, Figure 3), and additionally, hydroxylated and biotin-only surfaces were also prepared.

The PEG-biotin:PEG functionalization process was based on Soteropulos *et al.* [35]. In the first step, the silica surface was treated with piranha solution or oxygen plasma to populate the surface with terminal hydroxyl groups. Then, PEG is attached to the hydroxylated surface using a mixture of silane-PEG (2-[methoxy (polyethyleneoxy)$_{6-9}$propyl] trimethoxysilane, MPEOPS, MW = 460–590, purity >90%, Gelest) and silane-PEG-biotin (600 Da, Nanocs) (Figure 4). The applied ratios of PEG-biotin to PEG were 1:1, 1:2, and 1:3, respectively, with the 1:1 ratio equivalent to the same PEG density in solution as the Soteropulos study using MPEOPS. Afterwards, toluene (Certified ACS, ⩾99.5%, Fisher Scientific, f.w. 92.14), ethanol (Fisher Scientific, 95%) and deionized, distilled (DDI) water (Fisher Scientific, f.w. 18.02) were successively used to rinse the microspheres to remove physically adsorbed material.

	Non-specific protein
	Texas Red
	Avidin
	Biotin
	PEG

Figure 2. Goal: the PEG-biotin:PEG coating of the microsphere should improve the specificity by repelling non-specific protein adsorption. Avidin has four binding sites for biotin (two on each side); it is possible for avidin to bind 1–4 of these sites when it interacts with the surface, providing the biotin sites are within an appropriate distance. We expect avidin to bind to 1–2 biotin molecules on the surface, and present the other two binding sites to the environment; however, it is possible for a planar conformation to occur where all four sites bind the biotin tethered to the surface.

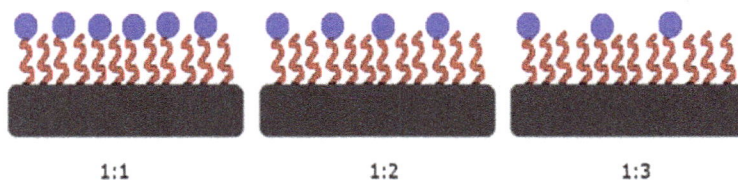

1:1   1:2   1:3

Figure 3. Two-dimensional schematic of the PEG-biotin:PEG ratios of 1:1, 1:2, and 1:3. The grey surface indicates the (100) silica-on-silicon wafer surfaces. The curved lines on the surface indicates the PEG coating, while the dark blue dots indicate the biotin molecules. Note that the amount of PEG-biotin in solution was kept constant, while the amount of PEG was varied.

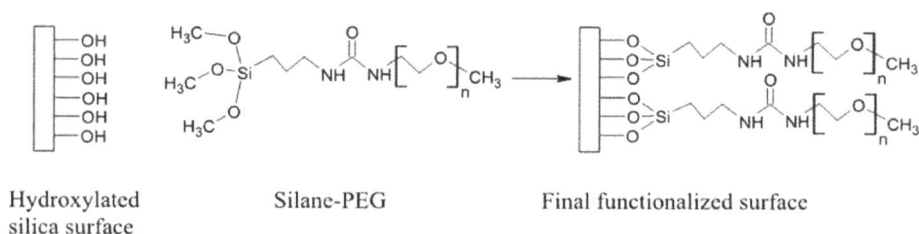

Hydroxylated   Silane-PEG   Final functionalized surface
silica surface

Figure 4. Functionalization process of silica microsphere's surface via a two-step covalent process. Adapted with permission from [35]. Copyright (2012) American Chemical Society. The silica surface was first populated with hydroxyl groups by exposing the device to piranha solution. The hydroxylated surface was then PEG-terminated and PEG-biotin terminated using silane coupling agents attached to PEG molecules via solvent-based, covalent grafting techniques.

The hydroxylated and biotin-only coatings were prepared using the protocols suggested in Hunt *et al.* [36]. The three-step conjugation is recyclable (Figure 5). First, the silica surface was terminated with hydroxyl groups with the use of oxygen plasma or piranha solution. Then, it was functionalized with 3-aminopropyltrimethoxy-silane (95%, Fisher) selectively in a vacuum desiccator for fifteen minutes. Afterwards, the surface was biotinylated with EZ-Link™ NHS-Biotin (Fisher) for thirty minutes and rinsed with DDI water. A stable amide bond was created from the NHS esters binding with the primary amines [36].

The thicknesses of the functionalized layers were then measured with a variable angle spectroscopic ellipsometer (VASE ellipsometer, J.A. Woollam, accurate to a fraction of a nm) as the initial thickness of PEG-biotin:PEG film, and were analyzed with the software Wvase32. Three randomly selected spots on each piece of wafer were measured in a wavelength range of 400–1000 nm by 10 nm increments, with angles varying from 65°–75° by 5° increments, and a dynamic averaging of 30. The measurements were taken under high accuracy mode and Isop + Depolarization sample type. To build a model fitting the data, a Cauchy layer with a calibrated thickness of 2031.336 nm was added to a 1 mm Si film (as defined in Si_jaw) to model the 2 μm silica layer of the silica-on-silicon wafer. A second Cauchy layer was added on the top to model the polymer layer. After the measurement, the wafers were then immersed in 1 mg/mL solutions of lysozyme (chicken egg white, Sigma Aldrich) or fibrinogen (bovine plasma, EMD Chemicals) in phosphate buffered saline (PBS, EMD Chemicals) for an hour, and subsequently rinsed with PBS and blown dry with nitrogen (Airgas, Ultra High Purity 5.0 Grade). The thickness measurement was taken again as the final thickness of the PEG-biotin:PEG film with non-specific adsorption, using the same settings as the initial measurement. The thickness change provides an indication of how much protein adsorption occurred on each surface.

The smaller the thickness changes, the greater the film's resistance to non-specific adsorption. The ratio minimizing protein adsorption could thus be selected.

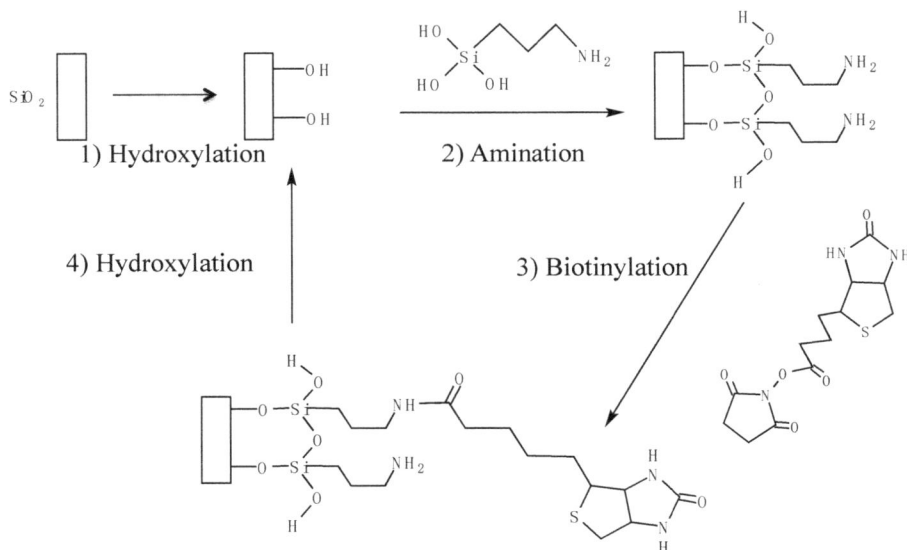

Figure 5. Overall reaction scheme for the biotinylation of silica surfaces (based on reference [36]). (1) Hydroxylation of the silica surface; (2) Amination of the hydroxylated surface via the silane coupling agent; (3) Biotinylation of the aminated surface via NHS ester chemistry; (4) Stripping of the surface via oxygen plasma, resulting in a hydroxylated surface.

Optical profilometry was performed to investigate the surface quality of the functionalized wafers and to compare it to the non-functionalized wafers. The nanometer-scale roughness data was obtained with an optical profiler (Veeco, WYKO NT 9109) in phase shift interferometry (PSI) mode and was analyzed with the software Vision 4.20. The data was presented with the arithmetic average of the absolute values of the roughness profile ordinates:

$$R_a = \frac{Z_i - Z_j}{n} \tag{1}$$

The Contact Angle Goniometer (ramé-hart Model 200 Standard, 200-F4, accurate to $0.1°$) measured the contact angle of a 2 μL DDI water drop on three randomly selected spots of each piece of wafer respectively, using the sessile drop method. The DROPimage Standard software recorded the instant contact angle for the PEG-biotin:PEG samples of three ratios, hydroxylated control, and biotin-only control. Young's equation describes the mechanical equilibrium of the three interfacial

108

tensions with the contact angle of a liquid drop [37]. The surface is considered hydrophilic when the contact angle is greater than 90°.

To examine if the PEG-biotin and PEG indeed deposited to the surface from the specific solution ratios, the fluorescence intensity of each sample was measured. Texas Red-Avidin was used as the fluorescent dye. Texas Red-Avidin (TR-Avidin) (Invitrogen, 2.5 mg/mL) was first diluted to 2 mg/mL with PBS and centrifuged to obtain the supernatant. The supernatant was then diluted to 10 μg/mL with PBS to incubate the functionalized wafers for 30 min in the dark, after which the wafers were rinsed with PBS and put on a cold hot plate, then warming to 40 °C for 2 min. The dye was attached to the surface due to the high affinity between biotin and avidin. Fluorescence microscopy was accomplished via an Olympus IX 70 system, using 20X magnification and the red filter. To find the best exposure time, a pseudo-color LUT (look-up table) was used to evaluate the brightness of the objects in the image during acquisition. In this LUT, the brightness of the image pixels was shown on an arbitrary scale from dark blue (black pixels with zero brightness) to white (saturated pixels with maximal brightness value 4096), as the exposure time was varied from short to long. In order to avoid saturated pixels, the exposure time was reduced until all "white" pixels were gone. This exposure time was then used for all samples. Three images, of randomly selected views, were acquired for each surface and the fluorescence intensity of four randomly selected regions of 300 nm × 300 nm in each of those images were measured with Metamorph software. Thus, twelve values of the intensity for each group (set of functionalization parameters) were obtained and then averaged during data analysis.

2.2. Device Fabrication, Functionalization, and Surface Characterization

Upon selecting the best ratio, the same functionalization process was applied to the silica microsphere optical microresonators. Silica microspheres with diameters of 200 μm were fabricated by melting the tip of a stripped, single-mode optical fiber (Single-mode, Newport F-SV) with CO_2 (Synrad) laser radiation at ~8% output power [38]. The microspheres were functionalized with PEG and PEG-biotin in a 1:1 ratio via the two-step covalent attachment process shown in Figure 2. To characterize the successful deposition of the functional groups to the microresonators, both optical microscopy and fluorescence microscopy were used, with the same procedures and parameters described above.

2.3. Device Characterization

To fabricate the tapered optical fiber that will be used as the waveguide, a hydrogen torch was used to heat the fiber while it was stretched across a two-axis stage controller, until it reached an average waist diameter of <700 nm [1]. The microsphere and the tapered fiber were coupled to each other, monitored by optical microscopy top-view and side-view cameras.

The Quality (Q) factor is a measure of the optical performance of the microresonator. It describes the deviation from the ideal resonator and is proportional to the confinement time, or the photon lifetime, of the circulating optical field confined by the microresonator [39]. A high value of the Q factor indicates longer photon lifetime, and more interactions of the optical field with the surrounding environment. The Q factor is a direct measure of the device sensitivity. For instance, microresonators with Q factors above 10^6 can be used for sensing single viruses [40]. In this study, the Q factor profile is recorded before and after each functionalization step, to ensure the optical microspheres' performance does not degrade due to the synthetic modifications.

To do this, light from a continuous wave (CW), tunable, diode laser with a center wavelength of 980 nm (New Focus, 6320H) is introduced to a single-mode optical fiber (Newport, F-SC). The optical field is then evanescently coupled, in the undercoupled regime, to the microresonator under investigation. The under-coupled regime is more favorable than the over-coupled and the critically coupled regime, due to the minimal extrinsic loss [41]. To control the coupling distance and to attain the desired regime, the fiber was fixed while the microsphere was moved via a fiber-holder (Thorlabs) attached to a three-axis nanopositioning stage (Optosigma). The device was monitored using side- and top-view cameras simultaneously (Figure 6).

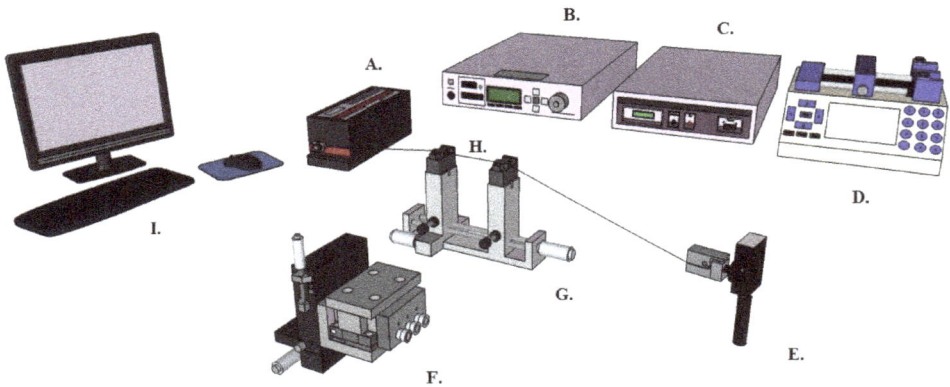

Figure 6. The experimental setup of device characterization. **A**—Laser **B**—Laser Controller **C**—Stage Controller **D**—Syringe Pump **E**—Photo Detector **F**—Nano-Positioning Stage **G**—Taper Holder **H**—Taper **I**—Computer. Light generated by a tunable diode laser propagates along an optical fiber. Once the Whispering Gallery Mode is excited upon coupling, the output signal is transferred to the detector and computer, and Q factor can be obtained automatically.

Once light was coupled into the resonator, the resonant frequency of the device could be detected via the switchable gain Si detector photodiode (Thorlabs, PDA36A). The resonance linewidth data was recorded using a digitizer/oscilloscope card directly integrated into the computer for automated data recording (NI, PCI-5153). The scan speed and direction of the laser was optimized to ensure that the resonance lineshape was not distorted. Since microspheres have multiple resonant frequencies, the frequency associated with the highest quality factor was used for the performance metrics. The resonant wavelength was recorded as λ; the peak data fitted with a Lorentzian function could give a full width at half maximum bandwidth (FWHM), denoted as Δλ. The mathematical expression for the Q factor can be summarized as: [42]

$$Q = \frac{\lambda}{\Delta\lambda} \tag{2}$$

2.4. Sensing

Non-specific adsorption sensing experiments, as well as avidin sensing experiments, were carried out through the use of an open-flow flow cell. The flow cell was constructed from glass slides and was fitted with metal tubing for injecting the buffer and analyte solutions as seen in Figure 7.

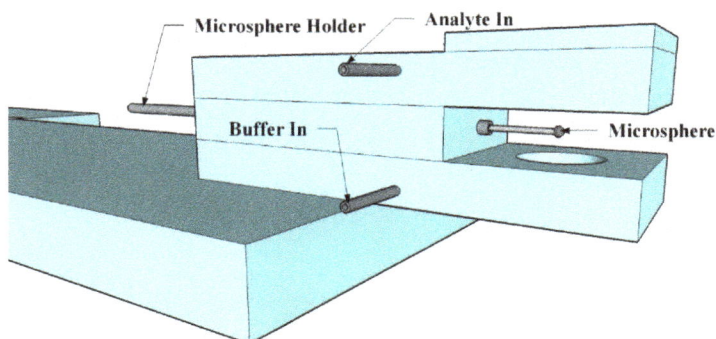

Figure 7. A model of the open-flow flow cell used for sensing experiments. This cell has injection ports for both the test molecule (avidin, lysozyme, or fibrinogen) and the PBS buffer solutions.

Sensing studies were performed using a custom LabVIEW program that located the lowest point on the oscilloscope graph of transmission voltage *versus* wavelength once the resonant wavelength with the highest quality factor was identified. The center wavelength (the lowest point on the graph) of this resonant wavelength was then tracked over time, resulting in wavelength shift data for each test molecule and concentration. For sensing experiments, the microresonator and the taper were in a coupled state. PBS buffer, which was used for the entirety of the study, was pumped up through the bottom of the flow cell using a syringe pump. The buffered microresonator-taper system was then allowed to reach equilibrium before any protein solution was added, thus allowing for the stable resonant wavelength of the system to be established. Previous literature has shown no significant resonant peak shift for control experiments completed using only PBS as the analyte [35]. A syringe pump was used to add various concentrations of the test molecule (avidin, lysozyme, or fibrinogen) into the flow cell. The PBS buffer was injected into the flow cell at a rate of 0.05 mL/min, while 0.04 mL of each test molecule solution was pumped into the flow cell at a rate of 0.03 mL/min. Solutions of lysozyme and fibrinogen, typical tests for non-specific adsorption, were created and tested at concentrations of 10, 100, 500, and 1000 µg/mL in PBS. Solutions containing avidin (Sigma Aldrich, from egg white, lyophilized powder) biotin also at concentrations of 10, 100, 500, and 1000 µg/mL, were then tested to analyze the effectiveness of the biotin recognition element and nonfouling surface that was fabricated on the microresonators.

To maintain consistency amongst the wavelength shifts and so that they were not confounded by using different microresonators, each sensing experiment was performed using the same coated microresonator. The sensing experiment consisted of evaluating fibrinogen, lysozyme, and avidin at each concentration, in that order, from the highest concentration to the lowest concentration, with a PBS rinse in between each concentration. The PBS rinse consisted of dipping the sphere in a solution of fresh PBS and placed on a rocker tray for 2 min. Then, the sphere was removed, dried, and re-rinsed using the same procedures. The entire sensing experiment was repeated two times, each time with a new, coated microresonator. Note also that the same size microresonators were used for all repeats to minimize the impact of a change in the microresonator size on the wavelength shift.

3. Results and Discussion

Characterization of Functionalized Wafers

As previously introduced, the (100) silica-on-silicon wafers were functionalized with three ratios of PEG-biotin to PEG: 1:1, 1:2, 1:3, with hydroxylated and biotin-modified surfaces serving as controls. To characterize the resistance of the coated surfaces to non-specific adsorption, the thickness change due to non-specific adsorption was calculated by subtracting the thickness of the film pre-adsorption from the thickness of the film post-adsorption. A smaller thickness increase indicated a thinner protein layer attached to the surface, which in turn suggested improvement in the coating's nonfouling characteristics.

The coatings of the as-functionalized wafers were measured with ellipsometry to be 9.7 ± 0.2 nm among the three ratios. The small standard deviation suggested that the thickness of the functionalized coatings on all the wafers, resulting from different PEG-biotin:PEG ratios, showed consistency. This measurement provided a uniform base measurement for each experiment group. After the adsorption, the thickness was recorded again. Figure 8 shows the thickness change for all three ratios and controls. It was observed that the non-specific adsorption layers on the PEG functionalized surfaces were thinner than those on the hydroxylated controls (lysozyme adsorption: 26.6 ± 0.7 nm; fibrinogen adsorption: 6.7 ± 0.1 nm), and similar to the biotin-only control (lysozyme adsorption: 2.3 ± 3 nm; fibrinogen-only adsorption: 1.4 ± 2 nm). The data further confirmed that our functionalization process indeed reduced non-specific adsorption relative to the hydroxylated control. Of greater interest is the biotin-only control (surfaces functionalized according to Figure 4), which suggests that the presence of the silane-biotin linker significantly reduces non-specific adsorption, without the need for additional nonfouling elements. However, the data from this control had a high standard deviation in comparison to the PEG-biotin:PEG surfaces, possibly due to the controls' lower uniformity. Interestingly, all three ratios of PEG-biotin:PEG reduced non-specific adsorption in approximately equivalent amounts, although we had hypothesized that the increased density of the PEG nonfouling agents would result in a significant decrease in non-specific adsorption. This means that the lowest PEG density of the 1:1 ratio is enough to reduce the non-specific adsorption, and more PEG is not needed to reduce it further. Therefore, the ratio 1:1 was selected for application to the three-dimensional optical microresonator, which minimizes the amount of PEG needed for the functionalization. This ratio should also have the highest relative amount of PEG-biotin immobilized on the surface, leading to an increased detection capacity. It is possible that a lower PEG density would work as well as 1:1; however, our previous work functionalizing with only PEG showed that this grafting density performed best at preventing non-specific adsorption [35].

The wafers that were exposed to adsorption were also examined by optical profilometry in order to determine their surface roughness parameters. Optical profilometry was performed in PSI mode to examine the surface quality of the wafers after non-specific protein (lyzosome) adsorption. The roughness parameters were presented as the arithmetic average of the absolute values of the roughness profile ordinates. The roughness of pre-functionalization surfaces were around 1.33 nm, while that of all the post-functionalization surfaces were around 1.5 nm (Figure 9). This indicates that the post-adsorption surfaces were still very smooth.

To study the physiochemical property change of the wafers after adsorption, contact angle was measured using the sessile drop method. As Table 1 shows, both functionalized and control surfaces became less hydrophilic after the adsorption. Before non-specific adsorption, PEG-biotin:PEG coating and the molecular weight of PEG, which is around 500, both make the surface exhibit a hydrophilic character. In addition, biotin also tends to be hydrophilic due to the presence of thiol and aldehyde groups. After non-specific adsorption, the hydrophilic groups were blocked by the adsorbed protein and tend to be hydrophobic. Therefore, the less hydrophobic (or more hydrophilic) a surface is following adsorption, the less non-specific adsorption has occurred.

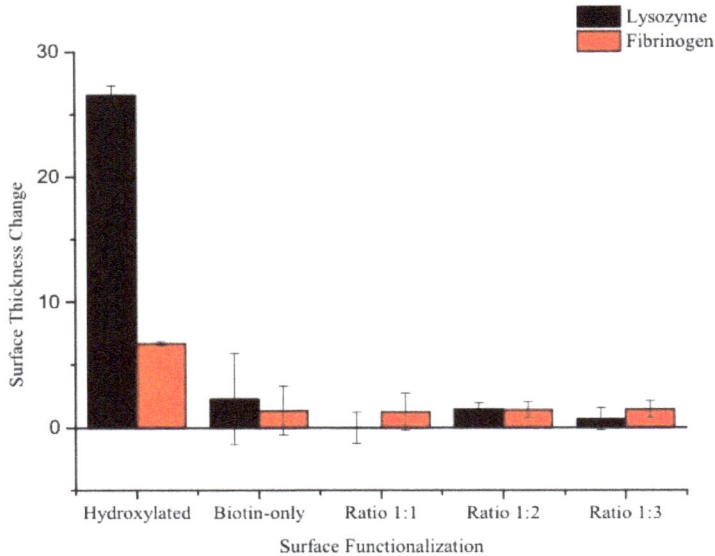

Figure 8. Mean thickness change ($\pm$standard deviation, nm) of functionalized wafers after adsorption. Three spots were measured on each wafer with ellipsometry, and three wafers were examined for each group. The values correspond to the amount of non-specific protein adsorbed onto the functionalized wafers.

For example, we see a dramatic increase in contact angle of the hydroxylated control upon adsorption, which indicates the hydroxylated surface changes to be very hydrophobic. This is consistent with the ellipsometry result that the hydroxylated control shows the largest thickness increase, and this further supports the conclusion that it has the highest levels of non-specific adsorption. The other samples had lower degrees of change in the contact angle and again the results for the ratios of PEG-biotin:PEG suggested that they all performed reasonably similarly, leading to the implication that only the 1:1 ratio needed to be evaluated (the lowest density of PEG). Interestingly, the biotin-only control surface, functionalized according to Figure 4, appeared to have similar resistance to non-specific adsorption, although this was unexpected.

114

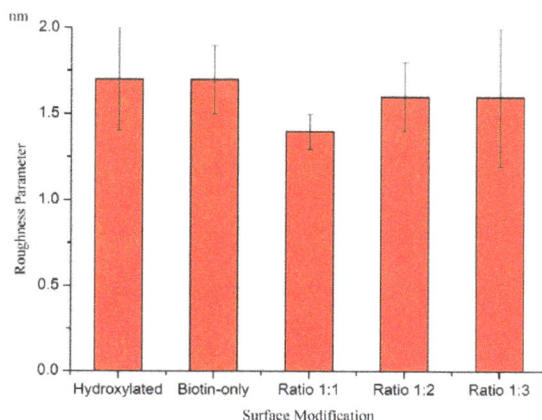

Figure 9. Mean Roughness Parameters (±standard deviation) of functionalized wafers after lysozyme adsorption. Three spots were measured on each wafer with optical profilometry, and three wafers were examined for each group. The addition of PEG to the surface appears to slightly reduce the overall roughness, although the initial surfaces are very smooth.

Table 1. Contact Angle Measurement before and after non-specific protein (Lysozyme and Fibrinogen) adsorption; the hydroxylated control experiences the largest increase in contact angle after adsorption, similar to the ellipsometry data, indicating that it experiences high adsorption, as expected.

	Contact Angle Measurement		
	Pre-Adsorption	Post-Adsorption (L)	Post-Adsorption (F)
Hydroxylated	65.8 ± 5.0	129.2 ± 5.7	90.3 ± 3.5
Biotin-only	77.0 ± 5.7	73.1 ± 4.6	89.2 ± 4.6
Ratio 1:1	62.2 ± 2.2	74.0 ± 2.7	81.3 ± 4.1
Ratio 1:2	63.2 ± 0.8	78.5 ± 5.5	80.5 ± 3.3
Ratio 1:3	65.8 ± 2.9	90.3 ± 9.3	86.5 ± 0.5

In addition, another group of wafers were functionalized with the same procedure and then were incubated in a solution of the fluorescent dye, Texas Red-Avidin, as previously described. We would expect to see a decrease in the level of fluorescence as the PEG-biotin:PEG ratio is decreased (increasing PEG) because the competitive binding levels should be relative to the solution levels as both the PEG-biotin and PEG molecules bind with identical mechanisms to the surface. Therefore, performing fluorescent intensity measurements provides information regarding the actual amount of biotin within the nonfouling surfaces. Figure 10 shows the fluorescence intensity measured with the Olympus IX 70. The data has been calibrated with a fluorescence-blank wafer subtracted from each group's intensity. The result shows that the hydroxylated surfaces (control) gave the lowest fluorescence intensity, while all of the other surfaces showed similar levels of fluorescence. Hydroxylated surfaces showed some intensity due to a small amount of avidin chemically and/or physically adsorbed to the surface (although similar amounts of washing to remove this were used with all reactions). When biotin probe molecules were attached to the surface (Biotin-only control and all ratios), the surfaces could successfully bind more avidin than the hydroxylated surfaces, and in turn showed more fluorescence. As the PEG-biotin:PEG ratio was changed from 1:1 to 1:3, we expect a reduction in fluorescence intensity, with slight variations present due to possible effects of competitive binding. However, we note that the nonfouling surfaces resulted in nearly identical levels of, or even slightly more, Texas Red—Avidin molecules immobilized compared with biotin-only control and no noticeable differences between the ratios tested. The potential for biotin to bind multiple avidin sites should not impact non-specific adsorption, however, just fluorescence. The fluorescence results suggest that there are sufficient biotin sites present across all 3 ratios to promote a pseudo-monolayer of avidin binding. Otherwise, we would expect a decrease in the fluoresence intensity as the amount of PEG is increased. This shows that the inclusion of the nonfouling chemistry does not decrease the surface's ability to bind with its target analyte. This further suggests there is no benefit in diluting the PEG-biotin concentration below the 1:1 ratio.

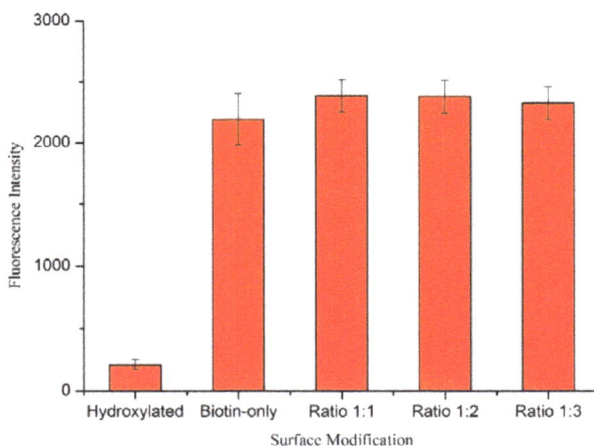

Figure 10. Mean fluorescence intensity (±standard deviation) of functionalized wafers. Five regions of 256 by 256 pixels were measured on each fluorescence image with an Olympus IX 70, and three images were taken for each group of wafers.

The selected PEG-biotin:PEG ratio of 1:1 was applied to the silica microspheres. The modified spheres were divided to two groups, one was for surface quality

116

examination, and the other was for sensing experiments. After labeling the coated spheres with Texas Red-Avidin, fluorescence microscopy was used to investigate the uniformity of the coating. Figure 11 shows an image of a functionalized microsphere. A robust and uniform coating can be observed. It indicates that the functionalization process was successful in coating microspheres with PEG-biotin:PEG solution.

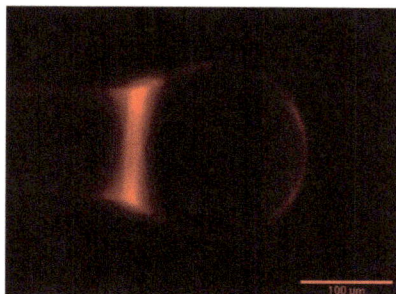

Figure 11. Representative fluorescence microscopy image of the microsphere functionalized by PEG-biotin:PEG from a ratio of 1:1. The functionalization process resulted in a uniform and smooth coverage on the surface.

The microresonators used in this study (three total, for repetitions of sensing experiments) were evaluated for their Q factor before and after coating. Figure 12 shows these values, as well as a representative resonance, fit with a Lorentzian function. Typically, the Q factor of the silica microspheres drop an order of magnitude after coating.

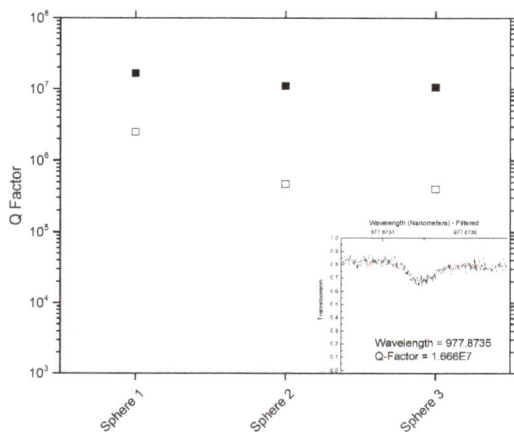

Figure 12. Quality factors of each microresonator used in the sensing study before (solid squares) and after (hollow squares) coating. Inset: a representative resonance (black line—data, red line—Lorentzian fit), showing a high quality factor device.

117

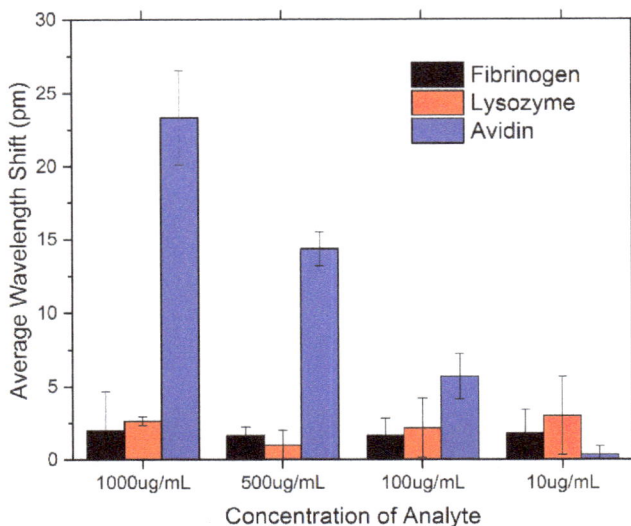

Figure 13. Average resonant wavelength shift of the PEG-biotin:PEG coated microresonators at concentrations of 1000, 500, 100, and 10 μg/mL for fibrinogen, lysozyme, and avidin.

Sensing results show that coating microresonators with PEG has effectively reduced the non-specific binding of both fibrinogen and lysozyme. Previous work has shown wavelength shifts in the range of 20–30 pm for bare (hydroxylated) microspheres [35]. Figure 13 shows the average resonant peak wavelength shift for all analytes tested at all concentrations, and Figure 14 highlights the specificity the coated microresonators have for avidin. We note that, for targeted sensing, a low limit of detection is necessary; however, for non-specific adsorption, a different metric should be evaluated to determine if the device performed well. In the case of non-specific adsorption, we typically want the surface to repel proteins, for example, at not just low concentrations but also high concentrations. Therefore, demonstrating that across a 100-fold increase in concentration, maintaining a statistically low level of background, non-specific adsorption, is important. Here, we show that not only can the surface "repel" or "block" non-specific adsorption at 10 μg/mL but also at the much more biologically relevant concentration of 1000 μg/mL.

118

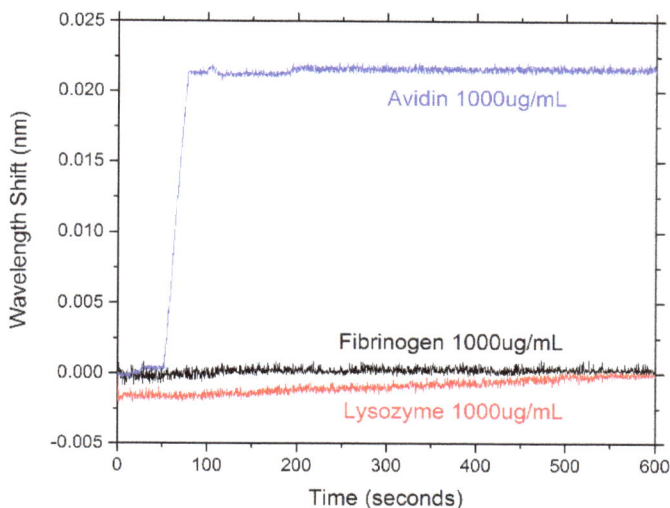

Figure 14. A representative example of the resonant wavelength shift of one of the PEG-biotin:PEG coated microresonators used in the study for different concentrations of fibrinogen, lysozyme, and biotin at the concentration of 1000 μg/mL. Fibrinogen and lysozyme show a minimal response, while avidin has a clearly defined wavelength shift that is approximately 20 pm.

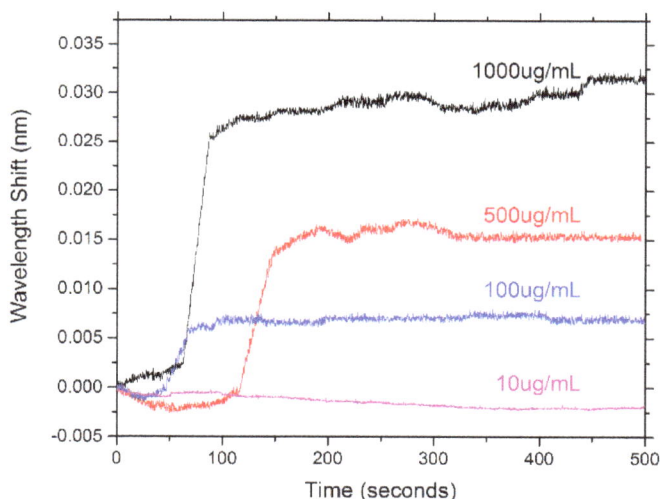

Figure 15. Representative response of one of the PEG-biotin:PEG coated microresonators for different concentrations of avidin. The wavelength shift declines as the concentration of avidin declines. Minimal response is seen at the concentration of 10 μg/mL.

119

Here, microresonators coated with PEG show an average shift in response to fibrinogen/lysozyme in the range of 0–5 pm. The sensing results with respect to avidin demonstrate that the coated microresonators can also recognize avidin in a concentration dependent manner. At avidin's highest concentration tested (1000 µg/mL), wavelength shifts in the range of 20–30 pm were seen. However, at avidin's lowest concentration tested (10 µg/mL), no significant wavelength shift was seen. Figure 15 illustrates the concentration dependent response of biotin for a given microresonator. The results in this study indicate that PEG-biotin:PEG coated microresonators may be able to effectively recognize avidin in a complex environment where non-specific proteins are present.

4. Conclusions/Outlook

In this study, we have demonstrated the nonfouling characteristics of PEG when coated on the surface of a Whispering Gallery Mode optical microresonator that is also functionalized with recognition elements (PEG-biotin). The reasoning for the use of this system, and, in particular, the use of biotin as a "recognition element", is that the biotin-avidin-biotin complex is frequently used as an intermediate sandwich complex when functionalizing surfaces of sensors. By first grafting biotin to the surface, then associating it with avidin, numerous biotin-labeled recognition elements can then be bound to the surface using the high affinity of avidin for biotin. The chemistry presented here, then, could be used as a general approach to reducing non-specific adsorption for targeted sensing using many different recognition elements. We evaluated the capability of different PEG-biotin:PEG ratios (1:1, 1:2, 1:3) in preventing non-specific adsorption, hypothesizing that the amount of exposed (PEG only) nonfouling elements would significantly impact the amount of non-specific adsorption. For each ratio chosen, the amount of the biotin recognition element (PEG-biotin) in solution was held constant while the amount of PEG nonfouling elements were increased. Fibrinogen and lysozyme were used to test for non-specific absorption to the PEG-biotin:PEG coated microresonators. The results showed that fibrinogen and lysozyme had minimal interactions with the coated wafers and microresonators, in comparison to the hydroxylated controls, and that increasing the PEG density on the surface did not significantly reduce non-specific adsorption beyond the 1:1 ratio. More interestingly, we found via the biotin-only control, which was functionalized with a simple silane coupling agent instead of PEG, yielded potentially comparable results to the PEG-biotin:PEG surfaces. This is surprising, as the silane coupling agent and the silane coupling agent used (aminopropyltrimethoxysilane, APTMS) are not known for their nonfouling properties in this sense. Avidin was then used to test for a specific interaction, and showed that the coated resonators were still capable of performing concentration-dependent detection. The combination of recognition and nonfouling elements should provide a means to increase the specificity of optical sensing by reducing the noise caused by non-specific adsorption. The covalently-bound nonfouling and recognition elements provide a means to increase the specificity of optical sensing by reducing the noise from non-specific adsorption. Due to the frequent use of biotin-avidin-biotin sandwich complexes in functionalizing sensor surfaces with biotin-labeled recognition elements, this chemistry could provide a common basis for creating a non-fouling surface capable of targeted detection. This should improve the ability of WGM optical biosensors to operate in complex environments, extending their application towards real-world detection.

Acknowledgments: Funding for this project included the University of Missouri Research Council, the Mizzou Alumni Association Richard Wallace Research Incentive Grant Award, and the University of Missouri College of Agriculture, Food and Natural Resources Research Interns Program. The authors would like to acknowledge Emily O'Brien and Emily Grayek for their useful discussions, and Mason Schellenberg for his assistance with the figures. Research was performed in part at the Christopher S. Bond Life Sciences Center and the Center for Micro/Nano Systems and Nanotechnology.

Author Contributions: F. Wang performed surface chemistry experiments. M. Anderson performed sensing experiments. F. Wang and M. Anderson wrote the paper. H.K. Hunt and M.T. Bernards designed the experiments, analyzed the results, and reviewed and edited the manuscript. All authors read and approved the manuscript.

Conflicts of Interest: The authors declare no conflict of interest.

References

1. Fan, X.; White, I.M.; Shopova, S.I.; Zhu, H.; Suter, J.D.; Sun, Y. Sensitive optical biosensors for unlabeled targets: A review. *Anal. Chim. Acta* **2008**, *620*, 8–26.
2. Borisov, S.M.; Wolfbeis, O.S. Optical biosensors. *Chem. Rev.* **2008**, *108*, 423–461.
3. Bentzen, E.L.; Tomlinson, I.D.; Mason, J.; Gresch, P.; Warnement, M.R.; Wright, D.; Sanders-Bush, E.; Blakely, R.; Rosenthal, S.J. Surface modification to reduce nonspecific binding of quantum dots in live cell assays. *Bioconjugate Chem.* **2005**, *16*, 1488–1494.
4. Vollmer, F.; Arnold, S. Whispering-gallery-mode biosensing: Label-free detection down to single molecules. *Nat. Methods* **2008**, *5*, 591–596.
5. Fang, Y. The development of label-free cellular assays for drug discovery. *Expert Opin. Drug Discov.* **2011**, *6*, 1285–1298.
6. O'Malley, S. Recent advances in label-free biosensors applications in protein biosynthesis and hts screening. In *Protein Biosynthesis*; Esterhouse, T.E., Petrinos, L.B., Eds.; Nova Science Publishers, Inc.: New York, NY, USA, 2008.
7. Homola, J. Surface plasmon resonance sensors for detection of chemical and biological species. *Chem. Rev.* **2008**, *108*, 462–493.
8. Situ, C.; Mooney, M.H.; Elliott, C.T.; Buijs, J. Advances in surface plasmon resonance biosensor technology towards high-throughput, food-safety analysis. *TrAC Trends Anal. Chem.* **2010**, *29*, 1305–1315.

9. Méjard, R.; Griesser, H.J.; Thierry, B. Optical biosensing for label-free cellular studies. *TrAC Trends Anal. Chem.* **2014**, *53*, 178–186.

10. Wilson, K.A.; Finch, C.A.; Anderson, P.; Vollmer, F.; Hickman, J.J. Whispering gallery mode biosensor quantification of fibronectin adsorption kinetics onto alkylsilane monolayers and interpretation of resultant cellular response. *Biomaterials* **2012**, *33*, 225–236.

11. Maleki, L.; Ilchenko, V. Optical Sensing Based on Whispering-Gallery-Mode Microcavity. U.S. Patents 6490039 B2, 3 December 2002.

12. Foreman, M.R.; Swaim, J.D.; Vollmer, F. Whispering gallery mode sensors. *Adv. Opt. Photonics* **2015**, *7*, 168–240.

13. Luchansky, M.S.; Bailey, R.C. High-Q Optical Sensors for Chemical and Biological Analysis. *Anal. Chem.* **2012**, *84*, 793–821.

14. Farca, G.; Shopova, S.I.; Elizondo, L.A.; Naweed, A.; Rosenberger, A.T. Evanescent-wave chemical sensing using WGM microresonators. In Proceedings of the 2004 Digest of the LEOS Summer Topical Meetings, San Diego, CA, USA, 28–30 June 2004.

15. Schweiger, G.; Horn, M. Effect of changes in size and index of refraction on the resonance wavelength of microspheres. *JOSA B* **2006**, *23*, 212–217.

16. Hu, J.; Sun, X.; Agarwal, A.; Kimerling, L.C. Design guidelines for optical resonator biochemical sensors. *JOSA B* **2009**, *26*, 1032–1041.

17. Vollmer, F.; Yang, L. Review Label-free detection with high-Q microcavities: A review of biosensing mechanisms for integrated devices. *Nanophotonics* **2012**, *1*, 267–291.

18. Armani, D.K.; Kippenberg, T.J.; Spillane, S.M.; Vahala, K.J. Ultra-high-Q toroid microcavity on a chip. *Nature* **2003**, *421*, 925–928.

19. Kuwata-Gonokami, M. Laser emission from dye-doped polystyrene microsphere. In Proceedings of the 1993 LEOS'93 Conference, San Jose, CA, USA, 15–18 November 1993; pp. 300–301.

20. Nöckel, J.U.; Stone, A.D. Ray and Wave Chaos in Asymmetric Resonant Optical Cavities. 1998, arXiv:chao-dyn/ 9806017. arXiv preprint. Available online: http://arxiv.org/ pdf/chao-dyn/9806017 (accessed on 2 January 2015).

21. Moon, H.-J.; Park, G.-W.; Lee, S.-B.; An, K.; Lee, J.-H. Laser oscillations of resonance modes in a thin gain-doped ring-type cylindrical microcavity. *Opt. Commun.* **2004**, *235*, 401–407.

22. Francois, A.; Himmelhaus, M. Optical biosensor based on whispering gallery mode excitations in clusters of microparticles. *Appl. Phys. Lett.* **2008**, *92*.

23. Merhari, L. *Hybrid Nanocomposites for Nanotechnology*; Springer: Berlin, Germany, 2009.

24. Yang, J.J.; Huang, M.; Yu, J.; Lan, Y.Z. Surface whispering-gallery mode. *EPL* **2011**, *96*.

25. Koyun, A.; Ahlatcolu, E.; Koca, Y.; Kara, S. Biosensors and their principles. In *A Roadmap of Biomedical Engineers and Milestones*; InTech: Rijeka, Croatia, Chapter 4.

26. Washburn, A.L.; Luchansky, M.S.; Bowman, A.L.; Bailey, R.C. Quantitative, Label-Free Detection of Five Protein Biomarkers Using Multiplexed Arrays of Silicon Photonic Microring Resonators. *Anal. Chem.* **2010**, *82*, 69–72.

27. Himmelhaus, M.; Krishnamoorthy, S.; Francois, A. Optical Sensors Based on Whispering Gallery Modes in Fluorescent Microbeads: Response to Specific Interactions. *Sensors* **2010**, *10*, 6257–6274.

28. Bog, U.; Brinkmann, F.; Wondimu, S.F.; Wienhold, T.; Kraemmer, S.; Koos, C.; Kalt, H.; Hirtz, M.; Fuchs, H.; Koeber, S.; *et al.* Densely Packed Microgoblet Laser Pairs for Cross-Referenced Biomolecular Detection. *Adv. Sci.* **2015**.

29. Zhu, H.Y.; Dale, P.S.; Caldwell, C.W.; Fan, X.D. Rapid and Label-Free Detection of Breast Cancer Biomarker CA15–3 in Clinical Human Serum Samples with Optofluidic Ring Resonator Sensors. *Anal. Chem.* **2009**, *81*, 9858–9865.

30. Obermeier, B.; Wurm, F.; Mangold, C.; Frey, H. Multifunctional poly(ethylene glycol)s. *Angew. Chem. Int. Ed.* **2011**, *50*, 7988–7997.

31. Golander, C.G.; Herron, J.N.; Lim, K.; Claesson, P.; Stenius, P.; Andrade, J.D.; Harris, J.M. *Poly(ethylene glycol) Chemistry: Biotechnical and Biomedical Applications*; Plenum Press: New York, NY, USA, 1992.

32. Zalipsky, S.; Harris, J.M. Introduction to chemistry and biological applications of poly(ethylene glycol). *Poly (ethyl. glycol)* **1997**, *680*, 1–13.

33. Chapman, R.G.; Ostuni, E.; Takayama, S.; Holmlin, R.E.; Yan, L.; Whitesides, G.M. Surveying for surfaces that resist the adsorption of proteins. *J. Am. Chem. Soc.* **2000**, *122*, 8303–8304.

34. Uchida, K.; Otsuka, H.; Kaneko, M.; Kataoka, K.; Nagasaki, Y. A reactive poly (ethylene glycol) layer to achieve specific surface plasmon resonance sensing with a high S/N ratio: The substantial role of a short underbrushed PEG layer in minimizing nonspecific adsorption. *Anal. Chem.* **2005**, *77*, 1075–1080.

35. Soteropulos, C.E.; Zurick, K.M.; Bernards, M.T.; Hunt, H.K. Tailoring the Protein Adsorption Properties of Whispering Gallery Mode Optical Biosensors. *Langmuir* **2012**, *28*, 15743–15750.

36. Hunt, H.K.; Soteropulos, C.; Armani, A.M. Bioconjugation strategies for microtoroidal optical resonators. *Sensors* **2010**, *10*, 9317–9336.

37. Smolders, C.A.; Duyvis, E.M. Contact angles; wetting and de-wetting of mercury: Part I. A critical examination of surface tension measurement by the sessile drop method. *Recl. Trav. Chim. Pays-Bas* **1961**, *80*, 635–649.

38. Soteropulos, C.E.; Hunt, H.K. Attaching biological probes to silica optical biosensors using silane coupling agents. *J. Vis. Exp.* **2012**.

39. Vahala, K.J. Optical microcavities. *Nature* **2003**, *424*, 839–846.

40. Vollmer, F.; Arnold, S.; Keng, D. Single virus detection from the reactive shift of a whispering gallery mode. *Proc. Natl. Acad. Sci. USA* **2008**, *105*, 20701–20704.

41. Kippenberg, T.J.; Spillane, S.M.; Armani, D.K.; Vahala, K.J. Fabrication and coupling to planar high-Q silica disk microcavities. *Appl. Phys. Lett.* **2003**, *83*, 797–799.

42. Gorodetsky, M.L.; Savchenkov, A.A.; Ilchenko, V.S. Ultimate Q of optical microsphere resonators. *Opt. Lett.* **1996**, *21*, 453–455.

Diamond Nanowires: A Novel Platform for Electrochemistry and Matrix-Free Mass Spectrometry

Sabine Szunerits, Yannick Coffinier and Rabah Boukherroub

Abstract: Over the last decades, carbon-based nanostructures have generated a huge interest from both fundamental and technological viewpoints owing to their physicochemical characteristics, markedly different from their corresponding bulk states. Among these nanostructured materials, carbon nanotubes (CNTs), and more recently graphene and its derivatives, hold a central position. The large amount of work devoted to these materials is driven not only by their unique mechanical and electrical properties, but also by the advances made in synthetic methods to produce these materials in large quantities with reasonably controllable morphologies. While much less studied than CNTs and graphene, diamond nanowires, the diamond analogue of CNTs, hold promise for several important applications. Diamond nanowires display several advantages such as chemical inertness, high mechanical strength, high thermal and electrical conductivity, together with proven biocompatibility and existence of various strategies to functionalize their surface. The unique physicochemical properties of diamond nanowires have generated wide interest for their use as fillers in nanocomposites, as light detectors and emitters, as substrates for nanoelectronic devices, as tips for scanning probe microscopy as well as for sensing applications. In the past few years, studies on boron-doped diamond nanowires (BDD NWs) focused on increasing their electrochemical active surface area to achieve higher sensitivity and selectivity compared to planar diamond interfaces. The first part of the present review article will cover the promising applications of BDD NWS for label-free sensing. Then, the potential use of diamond nanowires as inorganic substrates for matrix-free laser desorption/ionization mass spectrometry, a powerful label-free approach for quantification and identification of small compounds, will be discussed.

Reprinted from *Sensors*. Cite as: Szunerits, S.; Coffinier, Y.; Boukherroub, R. Diamond Nanowires: A Novel Platform for Electrochemistry and Matrix-Free Mass Spectrometry. *Sensors* **2015**, *15*, 12573–12593.

1. Introduction

Diamond, a natural as well as a synthetic material, has captured researchers' attention since decades. From any list summarizing the specific material properties, diamond is often at the extreme [1]: crystalline diamond shows the highest atomic density of any bulk crystal, the highest bulk modulus and the highest thermal conductivity. Diamond, a wide band gap semiconductor, is optically transparent from the far infrared to the ultraviolet, making it an ideal candidate for optical applications. During the growth of diamond films using chemical vapor deposition (CVD) systems, dopants and impurities can be readily incorporated into the material, allowing for tuning its optical and electrical properties. Substantial progress has been made in this area using boron for p-type doping and heavily boron-doped diamond (BDD) films (B-doping levels > 10^{20} cm^{-3}) are now produced routinely for electrochemical investigations. The interest of BDD films for electrochemical sensing arises from the wide potential window and negligible capacitive current achieved as well as their stability required for use in *in vitro* biosensing applications [2–7]. Figure 1 summarizes some examples of BDD applications in electrochemical sensing.

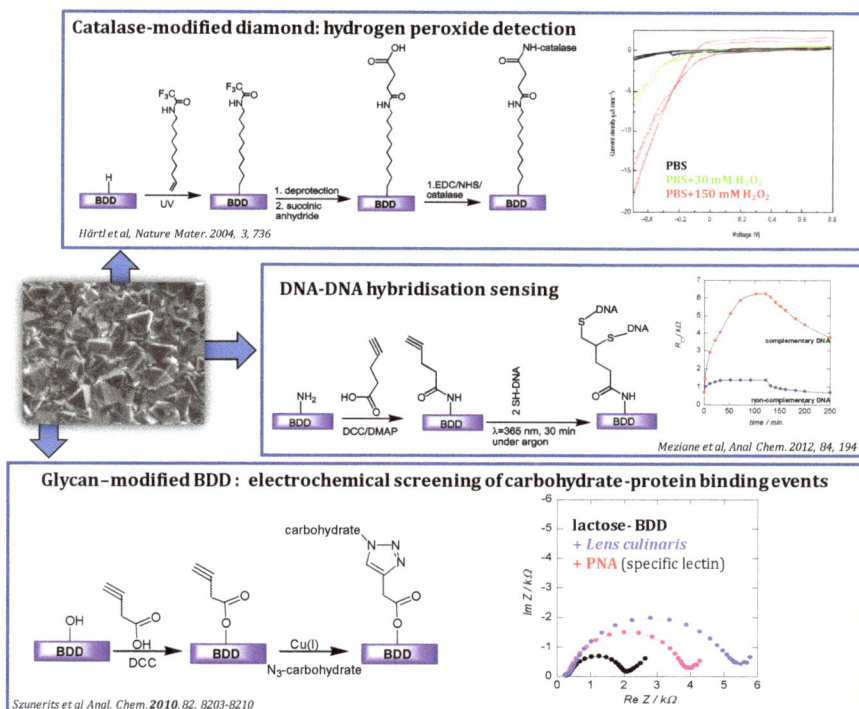

Figure 1. Selected examples of electrochemical sensing using BDD electrodes.

The attractiveness of diamond is that different morphologies and forms can be obtained from this sp^3 hybridized material (Figure 2). Indeed, modulation of the growth parameters results in microcrystalline to ultrananocrystalline CVD diamond films. Ultrananocrystalline films have the advantage of smooth surfaces, lower strain and improved fracture resistance. Such films are characterized by diamond domains that are ≈10 nm or less in size with thin sp^2-bonded boundaries.

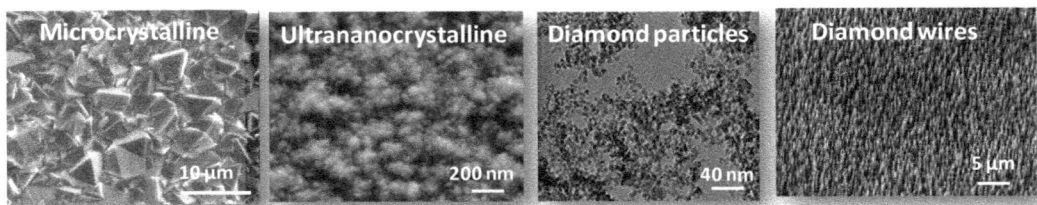

Figure 2. Different morphologies and forms of diamond.

Nanoscale diamond particles (also termed nanodiamonds, NDs) represent another interesting form widely explored for applications in drug delivery or medical diagnostics. The first records of the production of NDs date back to the 1960s, when a group of Soviet scientists discovered single crystals of cubic diamond particles in soot produced by detonating an oxygen-deficient TNT/hexogen composition in inert media without using any extra carbon source [8]. Since then, NDs have been rapidly gaining popularity in bioimaging and drug delivery applications, since colloidal suspensions of individual diamond particles with diameter of 4–10 nm are commercially available.

Diamond in the form of nanowires has to be added to this list. The use of diamond nanowires is believed to address positively issues related to improving the overall performance of sensors, including sensitivity and selectivity [9–12]. For a long time, the routine use of diamond nanowires was restricted, as no viable methods for their fabrication were available. The first attempt to synthesize diamond nanowires dates back to 1968 using a radiation heating unit developed from a super-high pressure Xenon lamp [13]. Only limited progress in this direction was made due to the difficulty of controlling the dimension of the diamond filament and lack of characterization facilities at that time, restricting further investigation of such diamond-based whiskers.

The intent of this review article is to make the reader more familiar with recent developments for the preparation of diamond nanostructures and their use for label-free sensing applications. The different synthetic methods can be classified into two main approaches: "top-down" and "bottom-up" approaches. Due to their high surface area, boron-doped diamond nanowires (BDD NWs) represent an interesting platform for electrochemical sensing as compared to planar BDD electrodes. This will be demonstrated in various examples of electrochemical sensing of different chemical/biological species using BDD NWs. Finally, the potential applications of diamond nanowires as inorganic matrix for surface-assisted laser desorption/ionization mass spectrometry (SALDI-MS) will be discussed in details.

2. Synthetic Routes of Diamond Nanowires

Reports on the fabrication of diamond structured surfaces with diameters as small as 25 μm and hundreds of microns in length date back to the 1960s [13]. However, it was only around the beginning of the 21st century that further attempts for the synthesis of diamond nanostructures were undertaken. The main approaches for the successful fabrication of diamond nanostructures are generally based on "top-down" and "bottom-up" processes.

2.1. Top-Down Approach

One of the initial attempts for a top-down synthesis of diamond nanostructures was reported by Shiomi [14]. Reactive ion etching (RIE) with oxygen plasma of CVD diamond film coated with a 400 nm thick aluminum layer resulted in columnar structures of approximately 300 nm in length and 10 nm in diameter (Figure 3A) [14]. The plasma-assisted RIE technology has since then been widely investigated for the top-down fabrication of diamond nanowires and pillars. Masuda *et al.* proposed, for example, the use of porous anodic aluminum oxide masks for the formation of diamond honeycomb films via oxygen plasma etching of CVD diamond films through the holes of porous alumina films (Figure 3B) [15]. They showed that the etching rate of alumina, compared to that of the diamond film, is negligible, resulting in the formation of honeycomb structures with high aspect ratios. Uniform holes with an average diameter of 70 nm and spacing of 100 nm with uniform depth of ~0.6 μm have been etched perpendicular to the diamond film surface, yielding an aspect ratio of ~9. An important fact is that this lithographic process was carried out in non-contact mode: the mask is merely placed on the substrate and therefore does not adhere to the substrate surface, unlike the situation with the resist-type masks commonly used in lithography. This makes the approach fast and easy to perform on different diamond interfaces.

Beside such masks, arrays of nanoparticles, seeded onto CVD grown diamond films, have been investigated by several groups as attractive alternatives. Aluminum [17], SiO_2 [18], gold [19], as well as diamond nanoparticles [11,20] proved to be useful etching masks (Figure 4A). Okuyama *et al.* used RIE with oxygen plasma through a two-dimensionally ordered SiO_2 particle array to form diamond cylinders [18]. The diameter and the length of the cylinders depend on the etching time and vary between 0.6–1 μm in diameter and 3–4 μm in length (Figure 4B) [18]. High-density and uniform diamond nanopillar arrays were obtained by Zou *et al.* by employing bias-assisted RIE in a hydrogen/argon plasma using gold nanoislands of 150 nm in diameter as etching masks [19]. The gold islands protect the underlying diamond from etching and sputtering; nanopillars with gold clusters at the tip are produced. Gold nanoparticles were indeed found to be one of the most suitable seeding masks as they are easy to disperse, resulting in single nanoparticles on the surface requiring no further processing step. The etch rate of the gold nanoparticles mask is 25 nm/min [21]. This yields good etch selectivity and diamond wires of 900 nm in height, and diameters from 275 nm at bottom to 310 nm at the top can be produced [21].

(A)   **(B)**

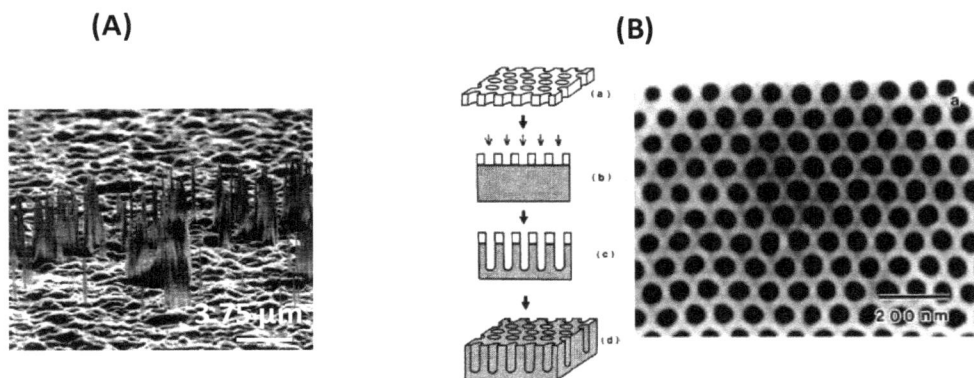

Figure 3. (**A**) SEM images of diamond nanowiskers formed on as-grown diamond films coated with 400 nm thick Al layer using reactive ion etching (RIE) with oxygen plasma (reprint with permission from [16]); (**B**) Formation of honeycomb diamond films (reprint with permission from [15]).

The use of nanodiamond (ND) particles as a hard mask for RIE of diamond was examined by research groups at the AIST (Japan) and IAF (Germany) [11,20,22,23]. Vertically aligned diamond nanowires were obtained using RIE in an O_2/CF_4 (97/3%) gas mixture for etching times of 10 s (Figure 4C). The length of the wires was limited by simultaneous etching of the ND particles mask with an etching rate of 10 Å·s^{-1}.

Mask-less top-down approaches have been recently proposed as alternatives [16,24–27] (Figure 5A). Such methods have the intrinsic advantage of being simple and straightforward, not requiring complicated processing steps such as mask deposition or template removal. Our group demonstrated that diamond nanowires can be easily prepared from highly boron-doped microcrystalline diamond thin films by RIE in an oxygen plasma (Figure 5A) [12,25–27]. The resulting nanowires are 1.4 ± 0.1 μm long with a tip and base radius of r_{tip} = 10 ± 5 nm and r_{base} = 40 ± 5 nm, respectively. The nanowires are about 140 times longer than aligned diamond nanowires prepared using diamond nanoparticles as a hard mask. X-ray photoelectron spectroscopy (XPS) analysis of the chemical composition of the diamond nanowires revealed that next to C_{1s} at 285 eV and O_{1s} at 532 eV, additional peaks at 402, 104 and 169 eV due respectively to N_{1s}, Si_{2p} and S_{2p} are present in the spectrum (Figure 5B). The latter elements are believed to originate from surface contamination during the RIE process. Indeed, the presence of SiO_x shell around the

BDD NWs was confirmed by HR-TEM analysis (Figure 5B). SiO_x was most likely deposited during the etching process due to sputtering of the substrate holder or the silicon wafer onto which the diamond film was deposited. A similar behavior was observed by Baik *et al.* when Mo sample holder was used [16]. The SiO_x deposits can be easily removed through immersion in HF aqueous solutions.

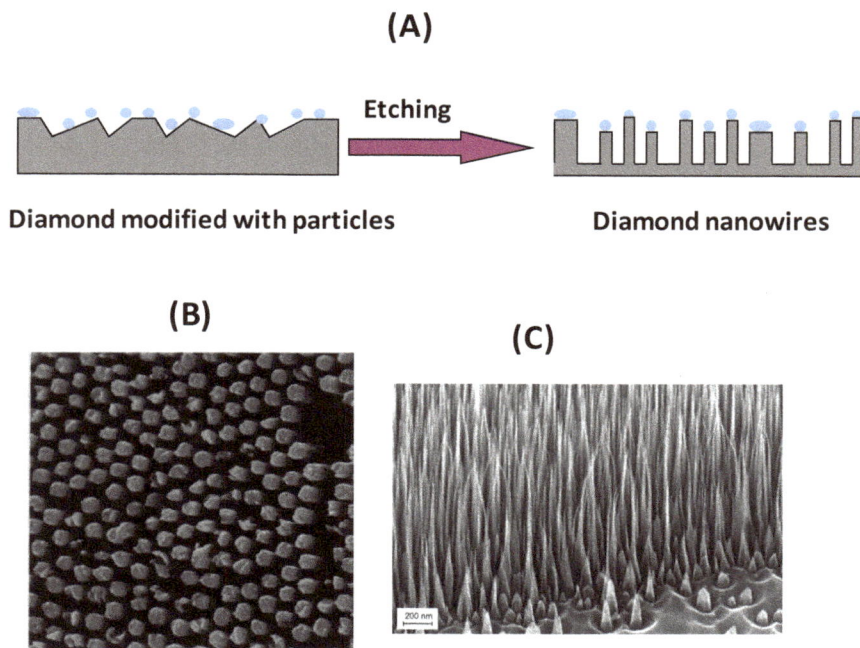

(A)

Diamond modified with particles → Etching → Diamond nanowires

(B)

(C)

Figure 4. (**A**) Top-down etching process of CVD diamond using seed particles; (**B**) SEM image of diamond cylinders obtained using oxygen reactive ion etching (RIE) for 60 min through a 1 μM SiO_2 particle array (reprint with permission from [18]); (**C**) SEM image of vertically aligned diamond nanowires using diamond nanoparticles as masks (with courtesy of C. Nebel).

(A)

Etching →

Diamond   Diamond nanowires

(B)

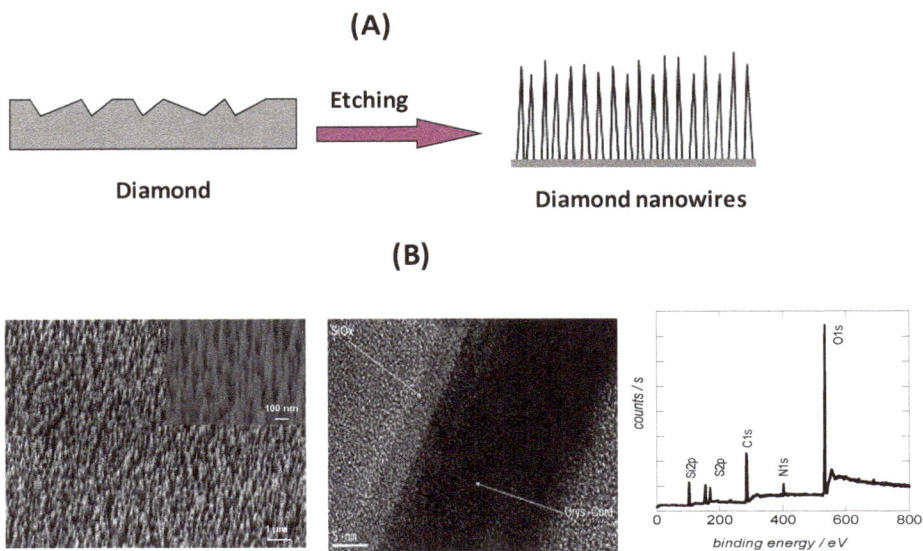

Figure 5. (**A**) Top-down etching process of CVD diamond without mask; (**B**) SEM and HR-TEM images, and XPS survey spectrum of boron-doped diamond nanowires synthesized through maskless technique (reprint with permission from [26]).

2.2. Bottom-Up Approach

As the formation of diamond nanowires by the top-down approach is often accompanied with surface damage, whose impact is generally greater at the nanoscale due to the large surface-to-volume ratio, bottom-up procedures were also investigated in the literature. One of the first bottom-up approaches was described by Masuda *et al.* (Figure 6A) [28]. In this technique, microcrystalline diamond nanocylinders were grown on anodic oxide templates using microwave plasma-assisted CVD and 50 nm nanodiamond particles as seeds. The density of the obtained wires was as high as 4.6×10^8 cylinders/cm^2 with a wire length of about 5 μm and 300 nm in diameter (Figure 6A). Diamond nanorods of 8–10 nm in diameter and up to 200 nm in length coated with an amorphous carbon layer were grown along the (110) direction upon applying a prolonged hydrogen plasma to multi-walled carbon nanotubes (MWCNTs) (Figure 6B) [29]. The authors suggested that initial diamond nuclei can be formed at defect sites of the MWCNTs due to the presence of hydrogen. At high temperature (1000 K) in the presence of hydrogen, MWCNTs transform to amorphous material, where the nucleation of diamond phase is facilitated. The addition of N$_2$ into the growth mixture of ultrananocrystalline diamond was reported by Vlasov and co-workers to change the surface morphology to wire-like structures [30]. The addition of 25% of N_2 to the $Ar/CH_4/H_2$ gas mixture resulted in the formation of hybrid diamond-graphite nanowires with lengths up to a few hundred nanometers. This hybrid material consists of a single crystalline diamond core of 5–6 nm in diameter oriented along the (110) principal axis and graphitic shells of different thicknesses covering the core. Using a mixture of N_2 and CH_4 allowed also the growth of ultrathin diamond nanorods by microwave-assisted CVD [31]. The resulting nanorods exhibited a diameter as small as 2.1 nm, which is not only smaller than any other reported diamond nanostructures but also smaller than the theoretical value of energetically stable diamond nanorods. More recently, the synthesis of straight, thin and long diamond nanowires using atmospheric-pressure chemical vapor deposition was proposed [32,33]. The diamond nanowires showed a uniform diameter of 60–90 nm with over tens of micrometers in length. Spectroscopic analysis provided information that these nanowires are diamond with high crystallinity and high structural uniformity.

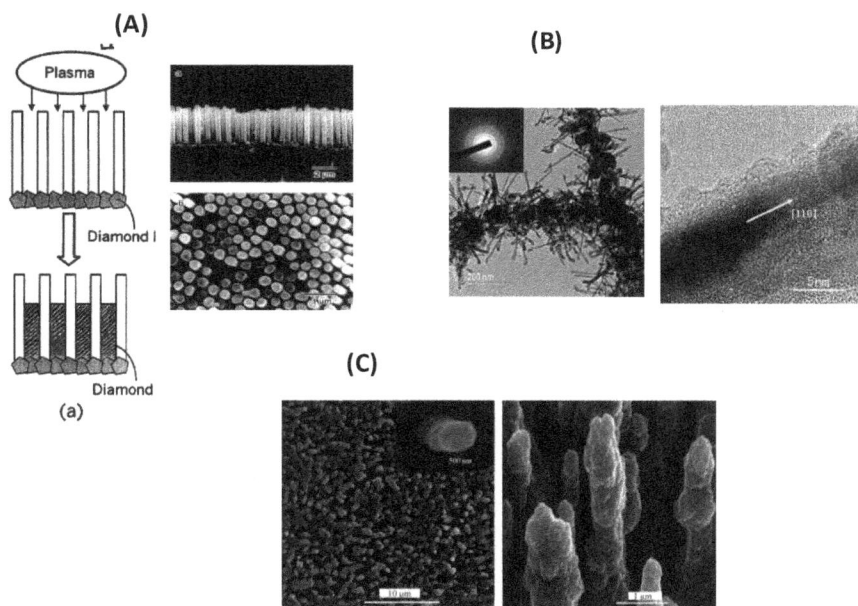

Figure 6. (**A**) Schematic diagram of the bottom-up fabrication of cylindrical diamond wires in porous alumina template (**left**) and the corresponding SEM images (reprint with permission from [28]); (**B**) TEM (**left**) and HR-TEM (**right**) images of diamond rods grown on carbon nanotubes (reprint with permission from [29]); (**C**) SEM image of diamond coated silicon nanowires (reprint with permission from [10]).

Post-coating of preformed silicon nanowires by diamond thin films has attracted considerable interest in the past five years for the preparation of core-shell nanowires [10,34,35]. Boron-doped diamond nanoforest electrode could be fabricated on pre-formed silicon nanowires as illustrated in Figure 6C [10]. Silicon nanowires (Si NWs) were formed according to a procedure reported by Peng *et al.* [36], and subsequently post-coated with boron-doped diamond thin films deposited by high frequency CVD technology. The coverage of the nanocrystalline diamond film is complete and continuous along the whole length (5 µm) of the Si NWs as seen in the SEM images in Figure 6C. A similar strategy was employed by Gao and co-workers to coat Si NWs with a 100 nm layer of nanocrystalline diamond by microwave-enhanced CVD [34]. The grafting of negatively-charged diamond nanoparticles over cationic polymer-coated nanometric patterns was proposed by Girard and co-workers as a bottom-up strategy towards 3D diamond nanostructures [35].

3. Applications of Diamond Nanowires

In general, the synthesis of diamond nanostructures has been advanced to a high level in a very short time span. The access to such nanostructures allowed finally the use of diamond nanowires for different applications ranging from solid-state electron emitters, high performance nano-electromechanical switches, probes for scanning probe microscopy and photonic systems, to the formation of superhydrophobic and oleophobic interfaces [15,21,26,31,33,37] (Figure 7). The use of boron-doped diamond nanowires has in particular found interest for electrochemical sensing [10–12,25,38–43] and as surface-assisted laser desorption/ionization (SALDI) matrix [27].

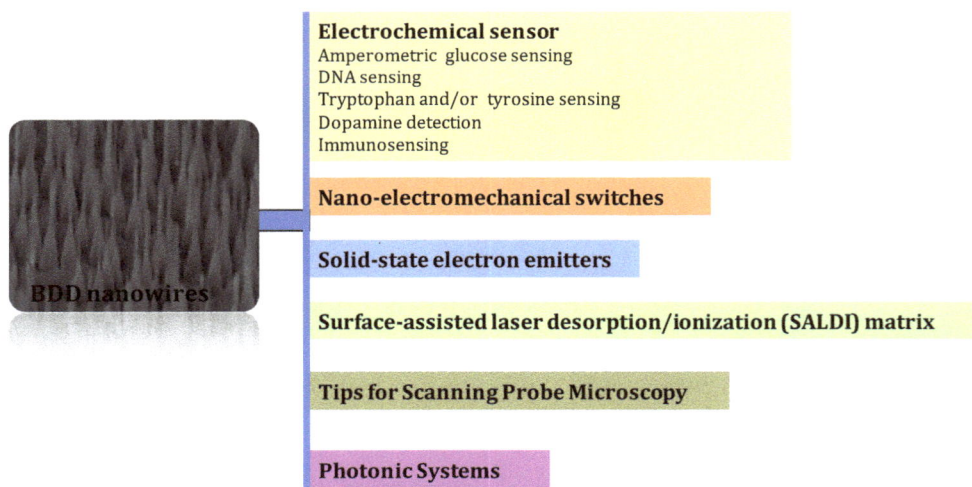

Electrochemical sensor
Amperometric glucose sensing
DNA sensing
Tryptophan and/or tyrosine sensing
Dopamine detection
Immunosensing

Nano-electromechanical switches

Solid-state electron emitters

Surface-assisted laser desorption/ionization (SALDI) matrix

Tips for Scanning Probe Microscopy

Photonic Systems

BDD nanowires

Figure 7. Applications of diamond nanowires.

3.1. Diamond Nanowires for Electrochemical Sensing

Diamond nanowires are among the fairly new but promising materials for chemical and biochemical sensing. The common theme of diamond sensors is that they convert biological or chemical information into an electrical signal, which can be measured accurately using a panel of electrochemical methods (e.g., cyclic voltammetry, differential pulse voltammetry, electrochemical impedance spectroscopy, etc.). As the technology required to create electrochemical biosensors is much cheaper than that required for fluorescence-based sensors, electrochemical sensors are dominating the analytical field. Their label-free character adds to their general value. While the conversion of a biological interaction to an electrical signal is attractive for sensors that are in continuous use or need to withstand harsh environments, so far, electrochemical sensors are in general several orders of magnitude less sensitive than the best fluorescence-based detection sensors.

Yang *et al.* were the first to demonstrate that the detection limit of electrochemical biosensors can be markedly improved if vertically aligned diamond nanowires are used [11]. DNA sensors were prepared through immobilization of single strand DNA probes onto diamond nanowires pre-functionalized with amine-terminated phenyl groups in an electrochemical functionalization step (Figure 8) [23,44]. The enhanced electrical field at the very end of the diamond tips resulted in a preferential DNA alignment at the tip rather than at the walls of the wires, increasing the probes' accessibility for interaction. This gave rise to optimized hybridization kinetics of complementary DNA (cDNA) and high sensitivity with a detection limit of 2 pM for cDNA as well as single-base mismatch discrimination.

Diamond nanowires electrodes allow also the direct electrochemical detection of glucose under strong basic conditions [10,12,45]. While almost no visible anodic peak for glucose oxidation was observed during the positive potential scan on a planar BDD electrode, a well-defined current response for glucose was obtained, for example, on BDD NWs electrodes of $\approx$3 μm in length with a diameter ranging from 10–50 nm [12]. The detection limit of this sensor was 60 μM (Figure 9). Such an improvement in glucose oxidation suggests that the Faradaic current of glucose oxidation depends strongly on the surface structure and porosity of the electrode, and the accessible surface area. The results reported by Nebel and co-workers showed that decoration of diamond nanopillar electrodes with Ni-nanoparticles improves the sensitivity of the sensor for glucose detection with a detection limit of 10 μM [45]. Diamond coated Si nanowires were investigated by Luo *et al.* for glucose sensing with an estimated detection limit of 0.2 μM and a sensitivity of 8.1 μA·mM^{-1}·cm^{-2} [10].

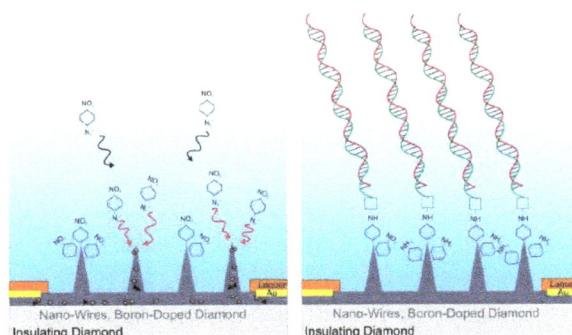

Figure 8. Diamond nanowires for DNA sensing: Preferential linking of phenyl aryl and DNA molecules to the tip of the wires (reprint with permission from [23,44]).

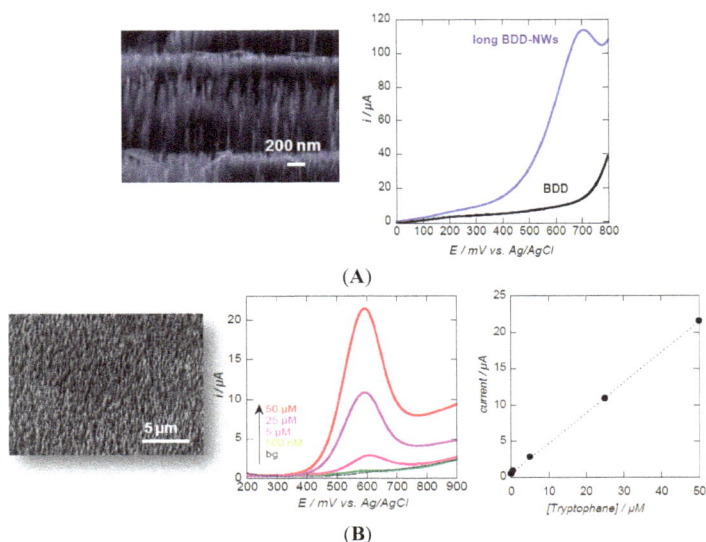

Figure 9. Diamond nanowires for enzyme-free glucose and tryptophan sensing: (**A**) SEM image of long diamond nanowires produced through mask-less RIE of CVD diamond films and linear sweep voltammogram recorded in 2 mM glucose solution (0.1 M NaOH) for BDD (black) and long BDD NWs (blue) (reprint with permission from [12]); (**B**) SEM image of short diamond nanowires together with differential pulse voltammograms for different concentrations of tryptophan and the corresponding calibration curve (reprint with permission from [25]).

Diamond nanowires are also adapted for the sensitive electrochemical detection of aromatic amino acids such as tryptophan and tyrosine, two important precursors of adrenaline, dopamine or melatonine [12,25]. A detection limit of 5×10^{-7} M was recorded for tryptophan on BDD NWs electrodes [25], being significantly lower than on planar microcrystalline BDD (1×10^{-5} M) [46]. The simultaneous detection of tryptophan and tyrosine by differential pulse voltammetry is also possible on BDD NWs electrodes, when the amount of tryptophan present in the mixture is not exceeding tryptophan/tyrosine $\leqslant 0.5$ [41]. The oxidation of other small molecules such as dopamine, uric acid and ascorbic acid was reported by Shalini *et al.* using nitrogen-doped diamond nanowire electrodes [47].

Electrochemical immunosensors based on chemically modified BDD NWs electrodes have been lately developed by our group [38–40]. Diamond nanowires immunosensors were constructed by coating diamond nanowires with functional conducting polymer films (e.g., carboxylic acid-terminated poly(pyrrole), copper ion (Cu^{2+}) chelation followed by linkage of histidine-tagged peptides [39]) (Figure 10A) or by electrochemical deposition of nickel nanoparticles (Ni-NPs) onto diamond nanowires followed by immobilization of biotin-tagged anti-IgG (Figure 10B) [38,40]. Post-coating of diamond nanowires with polymer films can be achieved by amperometrically biasing diamond nanowire electrodes at 1.2 V *vs.* Ag/AgCl in 3-(pyrrole) carboxylic acid solution [39]. Fine-tuning the charge allowed coating the wires rather than the formation of polymer films (PPA) in solution. Figure 10A shows SEM images of BDD NWs coated with carboxylic acid-terminated polypyrrole (PPA-BDD NWs) by varying the deposition time. At very low deposition charge (2 mC·cm^{-2}), the polymer started to form preferentially at the defect sites of the interface *i.e.*, in-between the BDD NWs. Increasing the deposition charge to 11 mC·cm^{-2} resulted in polymer coated BDD NWs, while large deposition charges (23 mC·cm^{-2} and higher) led to a loss of the wire structure in favor of continuous film formation. The available carboxylic groups of the poly(pyrrole) coated wire electrode allows their coordination with copper ion (Cu^{2+}), known to be specific binding sites for His-tagged analytes. Indeed, the affinity constant (K_A) of His-Tag-des-Arg^6-Bradykinine peptide to Cu^{2+} coordinated carboxyl-terminated diamond wires was determined as $K_A = (1.15 \pm 0.5) \times 10^6$ M^{-1}, higher than that determined in the absence of Cu^{2+} ($K_A = (0.31 \pm 0.5) \times 10^6$ M^{-1}). Concentrations as low as 10 nM resulted in R_{CT} shift of 50 ± 22 Ω on these interfaces, while on a planar BDD interface modified with carboxylic groups and chelated with Cu^{2+}, His-tagged peptide concentrations had to exceed 100 nM to cause a comparable shift.

The other approach consists of the deposition of nickel nanoparticles. Figure 10B(b) shows a representative SEM image of Ni NPs modified diamond nanowires (BDD NWs/Ni NPs) revealing particles of 20 ± 5 nm in diameter and a particle density of 150 Ni NPs/μm^2. The oxidation state of the deposited nickel was confirmed by the presence of a quasi-reversible redox peak at ≈ 0.47 V/Ag/AgCl in the cyclic voltammogram of the BDD NWs/Ni NPs electrodes in PBS (Figure 10B(c) assigned to $Ni(OH)_2$/NiOOH. The hydrated nickel hydroxide can be present in two crystalline forms: the hydrated α-Ni(OH)$_2$ and the anhydrous β-Ni(OH)$_2$, the latter being the more stable and preferentially formed by cycling in NaOH (0.1 M) [48]. Biotinylated anti-IgG was specifically linked to the Ni^{2+} particles and the change in the interfacial properties upon binding of different concentrations of IgG was detected using electrochemical impedance spectroscopy. A linear relation between ΔR_{ct} and IgG concentration in the range of 0.3–400 ng/mL was recorded (Figure 10A(d)) with a correlation coefficient of r = 0.9996 according to ΔR_{et} (kΩ) = 0.02 + 0.0451 × [IgG]. The detection limit of IgG was determined to be $\approx$0.3 ng/mL from five blank noise signals (95% confidence level). The advantage of this approach is the possibility of a controlled immobilization of biotinylated antibodies and antigens with the omission of avidin layers, influencing the electrochemical behavior of the electrical interface. As the affinity of the biotin-tag to the Ni NPs is weaker than avidin-biotin interaction, an easy regeneration of the interface through a simple ethylenediaminetetraacetic acid (EDTA) wash was achieved. This could allow in the future the immobilization of different antibodies using the same interface.

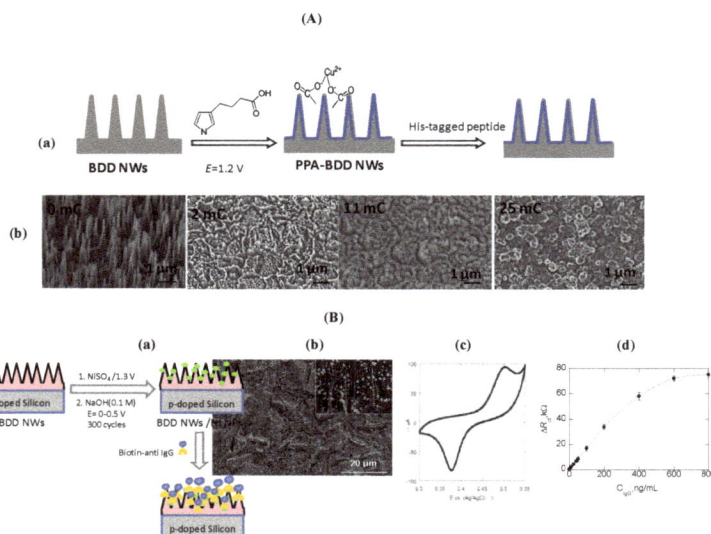

Figure 10. Diamond nanowires for immunosensing. (**A**) (**a**) fabrication method of polymer coated BDD NWs (PPA BDD NWs) with chelated Cu^{2+} ions and subsequently modified with histidine-terminated target; (**b**) SEM images of BDD NWs after electrochemical deposition of carboxylic acid-terminated poly(pyrrole) (100 mM) in TBATFB (0.1 M)/ acetonitrile solution at E = +1.2 V for different deposition charges (2, 11, 23 mC·cm^{-2}) (reprint with permission from [39]); (**B**) Fabrication method of Ni NPs modified diamond nanowires (**a**), SEM image of Ni NPs modified diamond nanowires (**b**), CV of Ni NPs-BDD NWs in 0.1 M NaOH (**c**), Calibration curve for IgG (**d**) (reprint with permission from [40]).

3.2. Diamond Nanowires for Matrix-Free Mass Spectrometry

Mass spectrometry (MS) is widely accepted as a 'gold-standard' method for the identification of chemicals or biological products. It is, nowadays, applied in highly diversified domains like those directly or indirectly tied to healthcare or regulation-driven demand such as drug development, diagnostics, food and environmental safety testing. Among the different methodologies, matrix-assisted laser desorption/ionization mass spectrometry (MALDI-MS), first introduced in 1988 by Hillenkampf and Karas [49], constitutes one of the soft ionization techniques that provides the nondestructive vaporization and ionization of analytes using UV-absorbing organic matrices. The utility of MALDI-MS for protein and peptide analysis lies in its ability to provide high accurate molecular-weight information on intact molecules. Acquiring optimum MALDI data depends, however, not only on the functional and structural properties of the analyte itself but also largely on the choice of suitable matrices. As the matrix is the medium by which the analyte is transported to the gaseous phase and provides the conditions that makes ionization possible, the matrix and the sample-matrix preparation procedure greatly influence the quality of MALDI mass spectra. Organic matrices such as 2,5-dihydroxylbenzoic acid (DHB), α-cyano-4-hydroxycinnamic acid (HCCA) or sinapic acid (SA) are commonly used. However, the choice of the matrix remains an empirical issue. While MALDI-MS has been successfully used to analyze large molecules [50–52], it has rarely been applied to low-molecular weight compounds (<500 Da) as a large number of matrix ions appear in the low-mass range.

Tanaka and co-workers proposed the use of a suspension of cobalt nanoparticles of 30 nm in size mixed with glycerol as inorganic matrix for the laser desorption/ionization of analytes [53]. Since then, several alternative inorganic matrices have been proposed as assisting material [54] in a process that was named surface-assisted laser desorption/ionization mass spectrometry (SALDI-MS) by Sunner *et al.* [55]. The basic principle of SALDI-MS analysis is schematically outlined in Figure 11A. Using inorganic species as the assisting material in MALDI is an alternative approach to avoid the problems arising in conventional MALDI analysis using organic compounds as matrix systems [56]. Nanostructured diamond-like carbon (DLC) coated targets were proposed by Bonn and co-workers as versatile platform for the analysis of amino acids, carbohydrates, lipids, peptides and other metabolites using laser desorption/ionization mass spectrometry [57]. A nanodiamond MALDI-MS support was demonstrated by Wei *et al.* to improve the ionization efficiency of samples [58]. The use of boron-doped diamond nanowires as an inorganic matrix for the D/I of peptides and small molecules, and their analysis by mass spectrometry with a very high sensitivity has been demonstrated by us [27]. To minimize droplet spreading on the matrix surface, the nanowires were chemically functionalized with octadecyltrichlorosilane (OTS) to reach a final water contact angle of 120 °C (Figure 11B). The sub-bandgap absorption under UV laser irradiation and the heat confinement inside the nanowires allowed LDI-MS of various compounds, most likely via a thermal mechanism (Figure 11B). One of the most used compounds to assess the LDI-MS performances of SALDI surfaces is verapamil, a calcium channel blocker. Detection of verapamil on silicon nanowires, prepared by the Metal-Assisted Chemical Etching (MACE) method, has shown a detection limit value of 5 fmol [59]. Walker *et al.* explored silicon nanopost arrays (NAPA) in combination with LDI-MS. The high ionization efficiencies enabled detection of ultratrace amounts of analytes (800 zmol of verapamil) within a dynamic range spanning up to four orders of magnitude [60]. For comparison, the detection limit value of verapamil on diamond nanowires was 200 zeptomoles, which is slightly better. Impressive results were obtained by Northen *et al.* by nanostructure-initiator mass spectrometry (NIMS). They demonstrated a detection limit value for verapamil of 700 ymol [61]. Although this technique is not a "matrix" *sensu stricto* (as MALDI), NIMS takes advantage of an "initiator" compound (e.g., bis(tridecafluoro-1,1,2,2-tetrahydrooctyl)tetramethyldisiloxane), which assists desorption and/or ionization of analyte molecules. Even though the limit of detection (LoD) that is needed depends on the targeted application, most of the time, the sensitivity race is not scientifically or even technically relevant. Other parameters should be taken into account for performant SALDI surfaces such as laser fluence, dynamic range, versatility, peptide discrimination and post-translational modification (PTMs) detection (proteomics field), cost, reproducibility.

The investigated diamond nanowires, undoped (UDD NWs) or boron-doped (BDD NWs), display antireflective properties permitting photons absorption at λ = 337 nm. However, on the mass spectrum of a peptide mixture obtained on BDD NWs interface, all peptides have been detected with relatively high signal intensity (Figure 11D). On the contrary, the same experiment performed on a planar nanocrystalline BDD, *i.e.*, the same interface without any RIE step process etching, no peaks were observed. Thus, the absence of peaks in the MS spectrum (Figure 11D) clearly indicates that the presence of nanostructures on the BDD substrate is mandatory for achieving the laser desorption/ionization (LDI) of biomolecules [27].

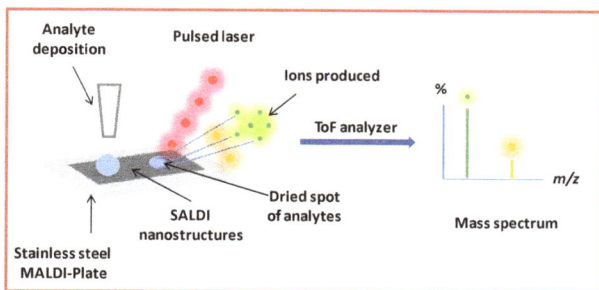

(A)

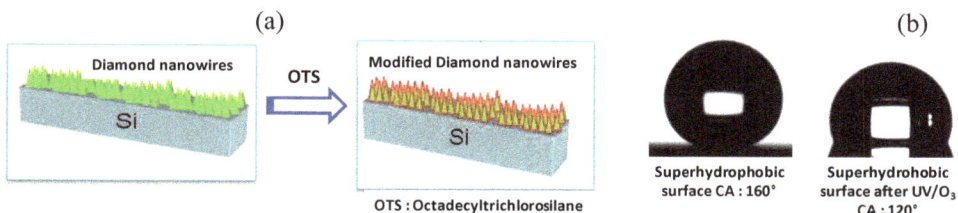

(B)

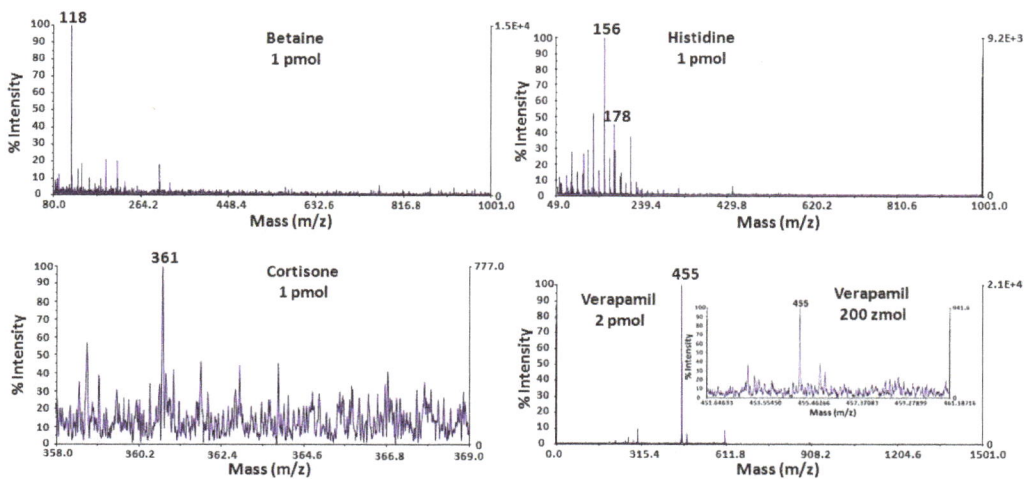

(C)

Figure 11. *Cont.*

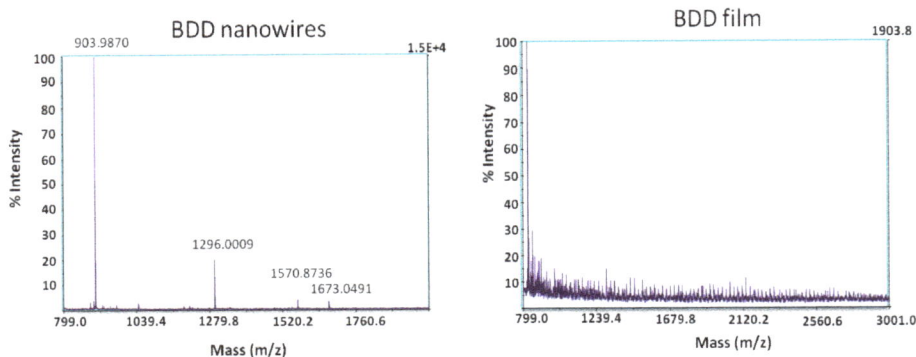

(D)

Figure 11. Diamond nanowires as inorganic matrices for SALDI. (**A**) Schematic presentation of the principle of surface-assisted laser desorption/ionization mass spectrometry (SALDI-MS) with analyte deposition on SALDI based nanostructures such as diamond nanowires; (**B**) OTS modified diamond nanowires (**a**) and contact angle values before and after UV/O$_3$ treatment (**b**); (**C**) LDI-MS spectra on BDD NWs of various compounds (Histidine, m/z 156, 1 pmol; Betaine m/z 118, 1 pmol; Cortisone m/z 361, 1 pmol and verapamil m/z 455, 2 pmol and 200 zmol (inset)); (**D**) LDI-MS detection of a peptide mixture ([Des-Arg^1]-bradykinin m/z 904 (50 fmol/μL), angiotensin I m/z 1296 (50 fmol/μL), [Glu^1]-fibrinopeptide B m/z 1570 (50 fmol/μL), neurotensin m/z 1673 (10 fmol/μL)) on BDD NWs and BDD films (reprint with permission from [27]).

3.3. Future Trend: Coupled Electrochemistry-Mass Spectrometry Analysis on Diamond Nanowires

Due to the great performance of BDD nanowires for electrochemical sensing and SALDI-MS, their combined use might provide a powerful quantitative and qualitative analysis platform. Electrochemistry coupled with mass spectrometry (EC/MS) allowed the identification of products or/and intermediates of electrochemical reactions, which is not only useful for elucidation of redox reaction mechanisms but also leads to many valuable bioanalytical applications [62,63]. The versatility of EC/MS stems from two facts. MS can serve as a sensitive and general detector for electrochemical compounds and can provide molecular weight information about an analyte of interest. In addition, tandem MS analysis can be used for structural determination based on ion dissociation. Electrochemical conversion, on the other hand, can improve analyte ionization and provides desired modification to the analyte prior to MS analysis. Attracted by the complementary nature of these two techniques, the coupling of EC and MS appears perfect and appealing. For

140

more than four decades, researchers have been engaged in coupling these two techniques [64–67].

A key challenge of the coupling is how to interface an electrochemical cell online with a mass spectrometer for efficiently ionizing the electrolyzed samples. Therefore, selecting an appropriate ionization method is critical. In this regard, the evolution of the coupled EC/MS technique followed the advances of ionization methods in the field of mass spectrometry. In 1971, the first EC/MS device was introduced by Bruckenstein and Gadde for *in situ* mass spectrometric determination of volatile electrode reaction products [64]. In the experiment, a Teflon membrane was placed between the porous electrode and the mass spectrometer ionization chamber so that volatile reaction products could penetrate through the membrane and be ionized by electron impact (EI) without the interference of the solvent. In 1984, differential electrochemical mass spectrometry (DEMS) was established for the online MS detection of volatile electrochemical products in real time. The total response time was less than 1 s [68]. Furthermore in DEMS, the product MS signal intensity was proportional to the Faradaic current as only the volatile compounds produced were transferred to the ionization chamber, from which quantitation could be achieved. Then, several methods of ionization have been used including thermospray (TS), fast atom bombardment (FAB), inductively coupled plasma (ICP), chemical ionization (CI), atmospheric pressure chemical ionization (APCI), atmospheric pressure photoionization (APPI) and electrospray ionization (ESI). This former combination (EC/ESI-MS) was the most widely used for many applications [62]. Desorption electrospray ionization (DESI) or nanoDESI and direct analysis in real time (DART) are other recent methods to perform ambient soft ionization with little or no sample preparation [69]. Applications of EC/MS are various and numerous and among them we can cite mechanistic elucidation of electrochemical reactions [70–73], mimicking of metabolic pathways [65], tagging of protein/peptide thiol groups using various electrochemical generated species to electrochemically enhance MS signals [74,75], pre-concentrate target via electrochemical deposition [66], and following protein/peptide cleavage and online MS analysis that is very notably useful in the proteomics field [76,77].

4. Conclusions

From the above discussion, it becomes clear that a large amount of effort has been devoted to the synthesis of diamond nanostructures to a point where they can be considered for device-oriented applications. The discoveries and research undertaken in the last years hope to trigger the development of diamond nanowire sensors for clinical diagnostic, environmental sensing and other applications at the interface between biology, physics and chemistry. However, the full spectrum of such nanostructures for other technological applications cannot be overseen.

Diamond coated silicon nanowires have been lately investigated for supercapacitor applications [34], bringing diamond nanostructures to the field of energy. A full and detailed understanding of the electrical and electrochemical properties of a single diamond nanowire might be of ultimate importance in the near future to foster further such developments.

The use of diamond nanowires is believed to have great potential for EC/MS. Indeed, coupling EC/MS using diamond nanowires will permit performing both electrochemistry and MS detection of ionized compounds achieved by either DESI (or nanoDESI) or the LDI ionization process.

Surface functionalization is required for almost any kind of sensing applications. Currently, the reported surface functionalization schemes of diamond wires are limited to some examples. Widening this area is thus one aspect that should be undertaken by research groups working in this field. The formation of superhydrophobic and oleophobic interfaces has, for example, been demonstrated to have impact on cell and bacteria adhesion [26]. An important aspect will be the determination of the influence of diamond doping levels and even wire length on SALDI results. A better understanding will help to optimize the technique and achieve highly reproducible and accurate results. This will make the approach of high interest for any laboratory.

Acknowledgments: Financial support from the Centre National de la Recherche Scientifique (CNRS), the Université Lille 1, the Nord Pas de Calais region and the Institut Universitaire de France (IUF) is gratefully acknowledged.

Conflicts of Interest: The authors declare no conflict of interest.

References

1. Nemanich, R.J.; Carlisle, J.A.; Hirata, A.; Haenen, K. CVD diamond—Research, applications, and challenges. *MRS Bull.* **2014**, *39*, 490–548.
2. Fortin, E.; Chane-Tune, J.; Mailley, P.; Szunerits, S.; Marcus, B.; Petit, J.P.; Mermoux, M.; Vieil, E. Nucleosides and ODN electrochemical detection onto boron-doped diamond electrodes. *Bioelectrochemistry* **2004**, *63*, 303–306.
3. McCreery, R.L. Advanced carbon electrode materials for molecular electrochemistry. *Chem. Rev.* **2008**, *108*, 2646–2687.
4. Hartl, A.; Schmich, E.; Garrido, J.A.; Hernando, J.; Catharino, S.C.R.; Walter, S.; Feulber, P.; Kromka, A.; Steinmuller, D.; Stutzmann, M. Protein-modified nanocrystalline diamond thin films for biosensor applications. *Nat. Mater.* **2004**, *3*, 736–742.
5. Ferro, S.; de Battist, A. The 5-V window of polarizability of fluorinated diamond electrodes in aqueous solutions. *Anal. Chem.* **2003**, *75*, 7040–7042.
6. Meziane, D.; Barras, A.; Kromka, A.; Houdkova, J.; Boukherroub, R.; Szunerits, S. Thiol-yne reaction on boron-doped diamond electrodes. *Anal. Chem.* **2012**, *84*, 194–200.

7. Szunerits, S.; Niedziółka-Jönsson, J.; Boukherroub, R.; Woisel, P.; Baumann, J.S.; Siriwardena, A. Label-free detection of lectins on carbohydrate-modified boron-doped diamond Surfaces. *Anal. Chem.* **2010**, *82*, 8203–8210.

8. Danilenko, V.V. On the history of the discovery of nanodiamond synthesis. *Phys. Solid State* **2004**, *46*, 595–599.

9. Szunerits, S.; Coffinier, Y.; Galopin, E.; Brenner, J.; Boukherroub, R. Preparation of boron-doped diamond nanowires and their application for sensitive electrochemical detection of tryptophan. *Electrochem. Commun.* **2010**, *12*, 438–441.

10. Luo, D.; Wu, L.; Zhi, J. Fabrication of boron-doped diamond nanorod forest electrodes and their application in nonenzymatic amperometric glucose sensing. *ACS Nano* **2009**, *3*, 2121–2128.

11. Yang, N.; Uetsuka, H.; Osawa, E.; Nebel, C.E. Vertically aligned diamond nanowires for DNA sensing. *Angew. Chem. Int. Ed.* **2008**, *47*, 5183–5185.

12. Wang, Q.; Subramanian, P.; Li, M.; Yeap, W.S.; Haenen, K.; Coffinier, Y.; Boukherroub, R.; Szunerits, S. Non-enzymatic glucose sensing on long and short diamond nanowires electrodes. *Electrochem. Commun.* **2013**, *34*, 286–290.

13. Derjaguin, B.V.; Fedoseev, D.V.; Lukyanovich, V.M.; Spitzin, B.V.; Ryabov, V.A.; Lavrentyev, A.V. Filamentary diamond crystals. *J. Cryst. Growth* **1968**, *2*, 380–384.

14. Shiomi, H. Reactive ion etching of diamond in O_2 and CF_4 plasma and fabrication of porous diamond for field emitter cathodes. *Jpn. J. Appl. Phys.* **1997**, *36*, 7745–7748.

15. Masuda, H.; Watanaba, M.; Yasui, K.; Tryk, D.; Rao, T.; Fujishima, A. Fabrication of a nanostructured diamond honeycomb film. *Adv. Mater.* **2000**, *12*, 444–447.

16. Baik, E.S.; Baik, Y.J.; Jeaon, D. Aligned diamond nanowhiskers. *J. Mater. Res.* **2000**, *15*, 923–926.

17. Ando, Y.; Nishibayashi, Y.; Sawaben, A. 'Nano-rods' of single crystalline diamond. *Diamond Relat. Mater.* **2004**, *13*, 633–637.

18. Okuyama, S.; Matsushita, S.I.; Fujishima, A. Periodic submicrocylinder diamond surfaces using two-dimensional fine particle arrays. *Langmuir* **2002**, *18*, 8282–8287.

19. Zou, Y.S.; Yang, T.; Zhang, W.J.; Chong, Y.M.; He, B.; Bello, I.; Lee, S.T. Fabrication of diamond nanopillar and their arrays. *Appl. Phys. Lett.* **2008**, *92*, 053105.

20. Yang, N.; Uetsuka, H.; Osawa, E.; Nebel, C.E. Vertically aligned nanowires from boron-doped diamond. *Nano Lett.* **2008**, *8*, 3572–3576.

21. Hausmann, B.J.M.; Khan, M.; Zhang, Y.; Bainec, T.M.; Martinick, K.; McCutcheon, M.; Hemmer, P.; Loncar, M. Fabrication of diamond nanowires for quantum information processing applications. *Diamond Relat. Mater.* **2010**, *19*, 621–629.

22. Smirnov, W.; Kriele, A.; Yang, N.; Nebel, C.F. Aligned diamond nano-wires: Fabrication and characterisation for advanced applications in bio and electrochemistry. *Diamond Relat. Mater.* **2009**, *18*, 186–189.

23. Nebel, C.E.; Yang, N.; Uetsuka, H.; Osawa, E.; Tokuda, N.; Williams, O. Diamond nano-wires, a new approach towards next generation electrochemical gene sensor platforms. *Diamond Relat. Mater.* **2009**, *18*, 910–917.

24. Zheng, W.W.; Hsieh, Y.H.; Chiu, Y.C.; Cai, S.J.; Cheng, C.L.; Chen, C. Organic functionalization of ultradispersed nanodiamond: Synthesis and applications. *J. Mater. Chem.* **2009**, *19*, 8432–8441.

25. Marcon, L.; Riquet, F.; Vicogne, D.; Szunerits, S.; Bodart, J.F.; Boukherroub, R. Cellular and *in vivo* toxicity of functionalized nanodiamond in Xenopus embryos. *J. Mater. Chem.* **2010**, *20*, 8064–8069.

26. Coffinier, Y.; Galopin, E.; Szunerits, S.; Boukherroub, R. Preparation of superhydrophobic and oleophobic diamond nanograss array. *J. Mater. Chem.* **2010**, *20*, 10671–10675.

27. Coffinier, Y.; Szunerits, S.; Drobecq, H.; Melnyk, O.; Boukherroub, R. Diamond nanowires for highly sensitive matrix-free mass spectrometry analysis of small molecules. *Nanoscale* **2012**, *4*, 231–238.

28. Masuda, H.; Yanagishita, T.; Yasui, K.; Nishio, K.; Yagi, I.; Rao, N.; Fujishima, A. Synthesis of well-aligned diamond nanocylinders. *Adv. Mater.* **2001**, *13*, 247–249.

29. Sun, L.T.; Gond, J.; Zhu, D.Z.; Zhu, Z.Y.; He, S. Diamond nanorods from carbon nanotubes. *Adv. Mater.* **2004**, *16*, 1849–1853.

30. Vlasov, I.I.; Lebedev, O.I.; Ralchenko, V.G.; Goovaerts, E.; Bertoni, G.; Tendeloo, G.V.; Konov, V.I. Hybrid diamond-graphite nanowires produced by microwave plasma chemical vapor deposition. *Adv. Mater.* **2007**, *19*, 4058–4062.

31. Shang, N.; Papakonstantinou, P.; Wang, P.; Zakharov, A.; Palnitkar, U.; Lin, I.N.; Chu, M.; Stamboulis, A. Self-assembled growth, microstructure, and field-emission high-performance of ultrathin diamond nanorods. *ACS Nano* **2009**, *3*, 1032–1038.

32. Hsu, C.H.; Cloutier, S.G.; Palefsky, S.; Xu, J. Synthesis of diamond nanowires using atmopsheric-pressure chemical vapor deposition. *Nano Lett.* **2010**, *10*, 3272–3276.

33. Hsu, C.H.; Xu, J. Diamond nanowire-a challenge from extremes. *Nanoscale* **2012**, *4*, 5293–5299.

34. Gao, F.; Lewes-Malandrakis, G.; Wolfer, M.T.; Muller-Sebert, W.; Gentile, P.; Aradilla, D.; Schubert, T.; Nebel, C.E. Diamond-coated silicon wires for supercapacitor applications in ionic liquids. *Diamond Relat. Mater.* **2015**, *51*, 1–6.

35. Girard, H.A.; Scorsone, E.; Saada, S.; Gesset, C.; Arnault, J.C.; Perruchas, S.; Rousseau, L.; David, S.; Pichot, V.; Spitzer, D.; Berganzo, P. Electrostatic grafting of diamond nanoparticles towards 3D diamond nanostructures. *Diamond Relat. Mater.* **2012**, *23*, 83–87.

36. Peng, K.Q.; Yan, Y.J.; Gao, S.P.; Zhu, J. Synthesis of large-area silicon nanowire arrays via self-assembling nanoelectrochemistry. *Adv. Mater.* **2002**, *14*, 1164–1167.

37. Babinec, T.M.; Hausmann, B.J.M.; Khan, M.; Zhang, Y.; Maze, J.R.; Hemmer, P.R.; Loncar, M. A diamond nanowire single-photon source. *Nature Nanotechnol.* **2010**, *5*, 195–199.

38. Subramanian, P.; Foord, J.; Steinmueller, D.; Coffinier, Y.; Boukherroub, R.; Szunerits, S. Diamond nanowires decorated with metallic nanoparticles: A novel electrical interface for the immobilization of histidinylated biomolecules. *Electrochim. Acta* **2013**, *110*, 4–8.

144

39. Subramanian, P.; Mazurenko, I.; Zaitsev, V.; Coffinier, Y.; Boukherroub, R.; Szunerits, S. Diamond nanowires modified with poly[3-(pyrrolyl)carboxylic acid] for the immobilization of histidine-tagged peptides. *Analyst* **2014**, *139*, 4343–4349.

40. Subramanian, P.; Motorina, A.; Yeap, W.S.; Haenen, K.; Coffinier, Y.; Zaitsev, V.; Niedziolka-Jonsson, J.; Boukherroub, R.; Szunerits, S. Impedimetric immunosensor based on diamond nanowires decorated with nickel nanoparticles. *Analyst* **2014**, *139*, 1726–1731.

41. Wang, Q.; Vasilescu, A.; Subramanian, P.; Vezeanu, A.; Andrei, V.; Coffinier, Y.; Li, M.; Boukherroub, R.; Szunerits, S. Simultaneous electrochemical detection of tryptophan and tyrosine using boron-doped diamond and diamond nanowires electrodes. *Electrochem. Commun.* **2013**, *35*, 84–87.

42. Yang, N.; Uetsuka, H.; Nebel, C.E. Biofunctionalization of vertically aligned diamond nanowires. *Adv. Funct. Mater.* **2009**, *19*, 887–893.

43. Yang, N.; Uetsuka, H.; Williams, O.A.; Osawa, E.; Tokuda, N.; Nebel, C.E. Vertically aligned diamond nanowires: Fabrication, characterization, and application for DNA sensing. *Phys. Stat. Sol. A* **2009**, *206*, 2048–2056.

44. Uetsuka, H.; Shin, D.; Tokuda, N.; Saeki, K.; Nebel, C.E. Electrochemical grafting of boron-doped single-crystalline chemical vapor deposition diamond with nitrophenyl molecules. *Langmuir* **2007**, *23*, 3466–3472.

45. Yang, N.; Smirnov, W.; Nebel, C.E. Three-dimensional electrochemical reactions on tip-coated diamond nanowires with nickel nanoparticles. *Electrochem. Commun.* **2013**, *27*, 89–91.

46. Zhao, G.; Qi, Y.; Tian, Y. Simultaneous and direct determination of tryptophan and tyrosine at boron-doped diamond electrode. *Electroanalysis* **2006**, *18*, 830–834.

47. Shalini, J.; Sankaran, K.J.; Dong, C.L.; Lee, C.Y.; Tai, N.H.; Lin, I.N. *In situ* detection of dopamine using nitrogen incorporated diamond nanowire electrode. *Nanoscale* **2013**, *5*, 1159–1167.

48. Toghill, K.E.; Xiao, L.; Stradiotto, N.R.; Compton, R.G. The determination of methanol using an electrolytically fabricated nickel microparticle modified boron-doped diamond electrode. *Electroanalysis* **2010**, *22*, 491–500.

49. Karas, M.; Hillenkamp, F. Laser desorption ionization of proteins with molecular masses exceeding 10,000 daltons. *Anal. Chem.* **1988**, *60*, 259–280.

50. Hanton, S.D. Mass spectrometry of polymers and polymer surfaces. *Chem. Rev.* **2001**, *101*, 527–569.

51. Knochenmuss, R.; Zenobi, R. MALDI ionization: The role of in-plume processes. *Chem. Rev.* **2003**, *103*, 441–452.

52. Li, L. *MALDI Mass Spectrometry for Synthetic Polymer Analysis (Chemical Analysis: A Series of Monographs on Analytical Chemistry and Its Applications)*; Wiley-VCH: Weinheim, Germany, 2009.

53. Tanaka, K.; Waki, H.; Ido, Y.; Akita, S.; Yoshida, Y.; Yoshida, T. Protein and polymer analyses up to m/z 100,000 by laser ionization time-of-flight mass spectrometry. *Rapid Commun. Mass Spectrom* **1988**, *2*, 151–153.

54. Arakawa, R.; Kawasaki, H. Functionalized nanoparticles and nanostructured surfaces for surface-assisted laser desorption/ionization mass spectrometry. *Anal. Sci.* **2010**, *26*, 1229–1240.

55. Sunner, J.; Dratz, E.; Chen, Y.C. Graphite surface-assisted laser desorption/ionization time-of-flight mass spectrometry of peptides and proteins from liquid solutions. *Anal. Chem.* **1995**, *67*, 4335–4342.

56. Qiao, L.; Liu, B.H.; Girault, H.H. Nanomaterial-assisted laser desorption ionization for mass spectrometry-based biomedical analysis. *Nanomedicine* **2010**, *5*, 1641–1652.

57. Najam-ul-Haq, M.; Rainer, M.; Huck, C.W.; Stecher, G.; Feuerstein, I.; Steinmueller, D.E.A. Chemically modified nano crystalline diamond layer as material enhanced laser desorption ionisation (MELDI) surface in protein profiling. *Curr. Nanosci.* **2006**, *2*, 1–7.

58. Wei, L.M.; Xue, Y.; Zhou, X.W.; Jin, H.; Shi, Q.; Lu, H.L.; Yang, P.Y. Nanodiamond MALDI support for enhancing the credibility of identifying proteins. *Talanta* **2008**, *74*, 1363–1370.

59. Piret, G.; Drobecq, H.; Coffinier, Y.; Melnyk, O.; Boukherroub, R. Matrix-free laser desorption/ionization mass spectrometry on silicon nanowire arrays prepared by chemical etching of crystalline silicon. *Langmuir* **2010**, *26*, 1354–1361.

60. Walker, B.N.; Stolee, J.A.; Vertes, A. Nanophotonic ionization for ultratrace and single-cell analysis by mass spectrometry. *Anal. Chem.* **2012**, *84*, 7756–7762.

61. Northen, T.R.; Yanes, O.; Northen, M.T.; Marrinucci, D.; Uritboonthai, W.; Apon, J.; Golledge, S.L.; Nordstrom, A.; Siuzdak, G. Clathrate nanostructures for mass spectrometry. *Nature* **2007**, *449*, 1033–1036.

62. Liu, P.; Lu, M.; Zheng, Q.; Zhang, Y.; Dewald, H.D.; Chen, H. Recent advances of electrochemical mass spectrometry. *Analyst* **2013**, *128*, 5519–5539.

63. Abonnenc, M.; Qiao, L.; Liu, B.H.; Girault, H.H. Electrocheical aspects of electrosptray and laser desorption/ionization for mass spectometry. *Annu. Rev. Anal. Chem.* **2010**, *3*, 231–254.

64. Bruckenstein, S.; Gadde, R.R. Use of a porous electrode for *in situ* mass spectrometric determination of volatile electrode reaction products. *J. Am. Chem. Soc.* **1971**, *93*, 793–794.

65. Jahn, S.; Karst, U. Electrochemistry coupled to (liquid chromatography/) mass spectrometry—Current state and future perspectives. *J. Chromatogr. A* **2012**, *1259*, 16–49.

66. Gutkin, V.; Gunand, J.; Lev, O. Electrochemical deposition-stripping analysis of molecules and proteins by online electrochemical flow cell/mass spectrometry. *Anal. Chem.* **2009**, *81*, 8396–8404.

67. Van Berkel, G.J.; Kertesz, V. Electrochemically initiated tagging of thiols using an electrospray ionization based liquid microjunction surface sampling probe two-electrode cell. *Rapid Commun. Mass Spectrom.* **2009**, *23*, 1380–1386.

68. Wolter, O.; Heitbaum, J. Differentiel electrochemical mass spectroscopy (DEMS)—A new method for the study of electrode processes. *Ber. Bunsenges. Phys. Chem.* **1984**, *88*, 2–6.

69. Takats, Z.; Wiseman, J.M.; Gologan, B.; Cooks, R.G. Mass spectrometry sampling under ambient conditions with desorption electrospray ionization. *Science* **2004**, *306*, 471–473.

70. Jurva, U.; Bissel, P.; Isin, E.M.; Igarashi, K.; Kuttab, S.; Castagnoli, N. Model electrochemical-mass spectrometric studies of the cytochrome P450-catalyzed oxidations of cyclic tertiary allylamines. *J. Am. Chem. Soc.* **2005**, *127*, 12368–12377.

71. Lu, W.; Xu, X.; Cole, R.B. On-line linear sweep voltammetry—electrospray mass spectrometry. *Anal. Chem.* **1997**, *69*, 2478–2484.

72. Zhang, T.; Palii, S.P.; Eyler, J.R.; Brajter-Toth, A. Enhancement of ionization efficiency by electrochemical reaction products in on-line electrochemistry/electrospray ionization Fourier transform ion cyclotron resonance mass spectrometry. *Anal. Chem.* **2002**, *74*, 1097–1103.

73. Zettersten, C.; Sjöberg, P.J.R.; Nyholm, L. Oxidation of 4-chloroaniline studied by on-line electrochemistry electrospray ionization mass spectrometry. *Anal.Chem.* **2009**, *81*, 5180–5187.

74. Roussel, C.; Rohner, T.C.; Jensen, H.; Girault, H.H. Mechanistic aspects of on-line electrochemical tagging of free L-cysteine residues during electrospray ionisation for mass spectrometry in protein analysis. *ChemPhysChem* **2003**, *4*, 200–206.

75. Rohner, T.C.; Rossier, J.S.; Girault, H.H. On-line electrochemcial tagging of cysteines in proteins during nanospray. *Electrochem. Commun.* **2002**, *74*, 695–700.

76. Roeser, J.; Permentier, H.P.; Bruins, A.P.; Bischoff, R. Electrochemical oxidation and cleavage of tyrosine- and tryptophan-containing tripeptides. *Anal. Chem.* **2010**, *820*, 7556–7565.

77. Roeser, J.; Permentier, H.P.; Bruins, A.P.; Bischoff, R. Oxidative protein labeling in mass-spectrometry-based proteomics. *Anal. Bioanal. Chem.* **2010**, *82*, 7556–7565.

147

A Label-Free Impedimetric DNA Sensor Based on a Nanoporous SnO$_2$ Film: Fabrication and Detection Performance

Minh Hai Le, Carmen Jimenez, Eric Chainet and Valerie Stambouli

Abstract: Nanoporous SnO$_2$ thin films were elaborated to serve as sensing electrodes for label-free DNA detection using electrochemical impedance spectroscopy (EIS). Films were deposited by an electrodeposition process (EDP). Then the non-Faradic EIS behaviour was thoroughly investigated during some different steps of functionalization up to DNA hybridization. The results have shown a systematic decrease of the impedance upon DNA hybridization. The impedance decrease is attributed to an enhanced penetration of ionic species within the film volume. Besides, the comparison of impedance variations upon DNA hybridization between the liquid and vapour phase processes for organosilane (APTES) grafting on the nanoporous SnO$_2$ films showed that vapour-phase method is more efficient. This is due to the fact that the vapour is more effective than the solution in penetrating the nanopores of the films. As a result, the DNA sensors built from vapour-treated silane layer exhibit a higher sensitivity than those produced from liquid-treated silane, in the range of tested target DNA concentration going to 10 nM. Finally, the impedance and fluorescence response signals strongly depend on the types of target DNA molecules, demonstrating a high selectivity of the process on nanoporous SnO$_2$ films.

Reprinted from *Sensors*. Cite as: Le, M.H.; Jimenez, C.; Chainet, E.; Stambouli, V. A Label-Free Impedimetric DNA Sensor Based on a Nanoporous SnO$_2$ Film: Fabrication and Detection Performance. *Sensors* **2015**, *15*, 10686–10704.

1. Introduction

Over the last decades, development of genosensors has increased significantly, as demonstrated by the large number of scientific publications on this topic [1]. Traditionally, DNA hybridization detection research has relied upon attachment of various labels to the molecules being studied. The common labels used in molecular biology studies to analyse DNA hybridization involve fluorescent dyes [2,3], redox active enzymes [4,5], magnetic particles [6] or different kinds of nanoparticles [7,8]. For example, the DNA target sequence is labelled with a suitable fluorescent tag. With the aid of a fluorescence microscope, fluorescence is observed at the place where complementary hybridization takes place [9]. Although these techniques are highly sensitive, label processes require extra time, expense,

148

sample handling [10]. Additionally, labels might, in some cases, interfere with the detection, the base-pairing interaction. The challenge is to develop simple, reliable and economical methods. Label-free strategies have emerged as potential methods for detecting DNA hybridization with lower cost and high sensitivity. Label-free techniques can provide direct information on target molecules in the form of changes in a physical bulk property of a sample. Basically, label-free DNA sensors rely on the modification of a given physical parameter of the supporting material (transducer), which is induced by DNA hybridization.

Electrochemical impedance spectroscopy (EIS) has received much attention recently for the DNA hybridization detection due to its ability to perform label-free detection. EIS can sensitively detect the change of the impedance of the electrode/electrolyte interface when the DNA target is captured by the probe. EIS measurements could be performed according either faradic or non-faradic process [10]. In the case of faradic impedance spectroscopy, the addition of a redox-active species, such as $[Fe(CN)_6]^{3-/4-}$ [11,12] or $[Ru(NH_3)_6]^{2+/3+}$ [13,14], to the bulk solution is required. Faradic EIS detection of DNA hybridization is generally based on the variation of the charge transfer resistance between the solution and the electrode surface [15]. On the other hand, no additional reagent is needed in the case of non-Faradic detection. Bio-modification of the electrode leads to the variation of either the capacitance of the double layer formed between the solution and the metal electrode surface or the capacitance located in the space charge layer at the sub-surface of semiconductive electrodes [16]. In this case, a sufficiently sensitive electrode material is strongly needed. Different kinds of sensitive materials for non-Faradic EIS DNA detection have been reported, including metals [17,18], conductive polymers [19–21] and semiconductors [22–29]. The latter can be divided into two categories including CMOS heterostructures [22–24] and single working electrodes [25–29].

Within this last category, our group pioneered to study the non-Faradic label-free detection of DNA hybridization based on semiconductive metal oxides as working electrodes. Dense and polycrystalline thin film electrodes constituted of $CdIn_2O_4$ [30,31] or pure/doped SnO_2 [32,33] were elaborated by the aerosol pyrolysis technique. The detection results first showed a systematic increase of the impedance upon DNA hybridization in agreement with the field effect. In particular, we evidenced the importance of the use of non-doped films to benefit from higher field effects. Elsewhere, the high chemical stability of SnO_2 films when dipped in saline solutions is an important criterion which led us to pursue further investigations with this metal oxide. In the following step, using an electrodeposition method, we elaborated working electrodes constituted of 1D monocrystalline nanopillars [34]. The dimensionality reduction of the SnO_2 electrode material from 2D thin film to 1D nanopillars allowed the surface/volume ratio of the electrode to increase and thus to benefit from an enhanced field effect. As a result, the increase of the impedance signal upon DNA hybridization was more important than in the case of 2D SnO_2 thin films. Our results showed that SnO_2-nanopillars-electrode provides a higher sensitivity over 2D-dense SnO_2 film electrode (97% $\pm$ 7% $vs.$ 50% $\pm$ 10%) for a DNA target concentration of 2.0 μM [34]. The limit of DNA detection was found in the nanomolar range, which we expect to improve in a future study by elaborating SnO_2 nanowires exhibiting a higher shape ratio. Presently, the idea is to reduce more the dimensionality of the electrode material down to 0D by elaborating nanoporous SnO_2 films constituted of SnO_2 nanoparticles and to investigate the resulting effect on the impedance signal upon DNA hybridization.

To this aim and for the first time to the best of our knowledge, in the present work, we investigated the possibility to fabricate impedimetric DNA biosensors based on nanoporous SnO_2 electrodes. As for SnO_2 nanopillar electrodes, the nanoporous SnO_2 film electrodes were prepared using an electrodeposition method which provides a simpler and less expensive route to synthesize the ceramic coatings over other methods [35]. The characteristics of the obtained films, including microstructure, morphology and electrochemical properties have been thoroughly investigated using SEM, TEM and EIS. Then, a functionalization process has been carried out in order to covalently graft single strand (ss) DNA probes onto the electrode film surface. This process is based on a silanization step that we have carried out either in liquid phase or in vapour phase. EIS was used to investigate the impedance behaviour after the main steps of the functionalization process, as well as after DNA hybridization. In parallel, the DNA hybridization detection on the SnO_2 nanoporous films was systematically checked using epifluorescence microscopy. Some performances of the sensors were also analysed, namely: sensitivity and selectivity.

The paper is organized as follows: we first present the results obtained for DNA hybridization when using the liquid phase silanization in the case of SnO_2 films with increasing thicknesses. Then we present the results obtained when using the vapour phase silanization. The comparison between these two steps will be conducted in term of impedance variation upon DNA hybridization. Finally, the obtained results help us to have a more complete view and understanding on the effect of the SnO_2 sensing electrode morphology and dimensionality on the response signals to non-faradic DNA detection.

2. Experimental Section

2.1. Nanoporous SnO_2 Film Deposition

The electrodeposition of SnO_2 thin films was carried out in a standard three-electrode electrochemical system using a computer-controlled potentiostat

EG&G 322. The electrolyte consisted of 20 mM $SnCl_2 \cdot 2H_2O$ (>99.99%, Sigma Aldrich, MO, USA), 100 mM $NaNO_3$ (>99%, Sigma Aldrich) and 75 mM HNO_3 (>65%, Sigma Aldrich) in Nanopure water. Commercial indium tin-oxide (ITO) coated glass substrates, purchased from Advanced Film Services Company (San Jose, CA, USA) were used as the working electrodes. The thickness of the ITO layer is 300 nm, with a sheet resistance of 10 Ω/square. These substrates were sonicated in the following sequence: 15 min in ethanol, 15 min in acetone and 15 min in isopropanol in order to remove all the impurities on the surface. Then, the ITO/glass substrate was installed into the cell vertically using a specific Teflon holder which controls the area of the working electrode exposed to the electrolyte 1 cm^2. A Pt wire and a commercial Ag/AgCl (KCl 3M) electrode were used as counter and reference electrodes, respectively. SnO_2 films were deposited on ITO substrates at potentiostatically a fixed potential of -1.0 V (*vs.* ref.).

Cathodic electrodeposition of SnO_2 film in nitrate solution comprises several steps [36]. First, in a strong oxidizing environment of nitric acid solution, the Sn^{2+} ions dissolved from tin dichloride are oxidized to Sn^{4+}. When the negative voltage is applied, nitrate ions are electrochemically reduced at the electrode surface leading to the generation of OH^- by Reaction (1). These formed OH^- ions then reacted with the Sn^{4+} ions coming from the bulk solution to deposit SnO_2 on the electrode surface according to Reaction (2).

$$NO_3^- + 2H^+ + 2e^- \rightarrow NO_2^- + 2OH^- \tag{1}$$

$$Sn^{4+} + 4OH^- \rightarrow Sn(OH)_4 \rightarrow SnO_2 + 2H_2O \tag{2}$$

Because the total charge density (Q) is proportional to the amount of NO_3^- electrochemically reduced to generate OH^- group at the electrode surface, Q relates to the amount of deposited SnO_2. As the result, the film thickness could be controlled by changing the value of Q. By increasing the Q values from 0.2 to 0.8 $C \cdot cm^{-2}$, SnO_2 films with increasing thickness were obtained.

2.2. Functionalization Process

The functionalization process of SnO_2 films leads to a covalent attachment of DNA. It is similar to the one we previously used for SnO_2 films and SnO_2 nanopillars [32,34]. Briefly, it consists of the following steps: the oxide film surface was first hydroxylated using an air/O_2 mixture plasma to create OH^- groups at the surface. These groups allowed covalent binding of a functional organosilane. Then a silanization step was accomplished by grafting of the 3-aminopropyltriethoxysilane (APTES). Both liquid-phase and vapour-phase procedures have been tested for APTES deposition on SnO_2 surface:

2.2.1. Liquid Phase Deposition

The samples were located into a solution containing 0.5 M of APTES (Sigma-Aldrich) in 95% absolute ethanol and 5% deionized water under agitation for a night. To remove the unbound silane, the samples were carefully rinsed with ethanol and then, with deionized water. This process was followed by curing the samples in an oven at 110 °C for 3 h.

2.2.2. Vapour Phase Deposition

The samples first were placed in a Teflon holder, which then was put into a glove bag. The next step was to draw out the air from the bag using a rotary pump and fill the bag with an argon gas. This step was repeated three times to make sure that the humidity in the bag is as low as about 5%. After 200 μL of APTES was delivered, the lid of the sample holder was closed tightly. This holder was kept at 82 °C for 1 h to cause the evaporation of APTES. To finish, the samples were rinsed carefully with absolute ethanol and deionized water to remove unreacted silane and cured in an oven at 110 °C for 1 h.

To facilitate strong covalent binding between the NH_2 termination of APTES and the 5'-NH_2 termination of the oligonucleotide, a cross linker molecule (10% glutaraldehyde solution in H_2O) was applied. 20-base pre-synthesized DNA probes were used (purchased from Biomers, Ulm, Germany). A standard-type probe sequence was chosen: 5'-NH_2-TTTTT GAT AAA CCC ACT CTA-3'. These DNA probes were diluted in a sodium phosphate solution 0.3 M/H_2O to a concentration of 10 μM. Two μL drops of this solution were manually applied on the sample surface and incubated for 2 h at room temperature. The probes were then reduced and stabilized using a $NaBH_4$ solution (0.1 M) which modifies the CH=N imine into a CH_2-NH amine bond and also deactivates the non-bonded CHO termination of the glutaraldehyde transforming them into CH_2-OH. The hybridization was carried out using DNA targets labeled with a Cy3 fluorescent dye. The DNA target solution was diluted in a hybridization buffer solution (NaCl: 0.5 M, PBS: 0.01 M) and spread throughout the sample surface. To minimize the experimental dilution errors, the DNA target solution was prepared once at 2 μM and was then diluted to the desired lower concentrations down to 10 nM. The samples were then placed into a hybridization chamber at 42 °C for 45 min. Finally, the samples are rinsed with saline-sodium citrate (SSC) buffer to remove all the unbound DNA targets from the surface and dried with nitrogen. In order to study the selectivity of the process, different types of DNA target have been used including complementary, non-complementary, 1- and 2-base mismatch as reported in Table 1.

Table 1. Sequences of the different types of DNA target.

Complementary	3' AC CTA TTT GGG TGA GAT AC-Cy3 5'
Non-complementary	3' AC TGG CGC AAT CAC TCT AC-Cy3 5'
1-base mismatch	3' AC CTA TTT G**C**G TGA GAT AC-Cy3 5'
2-base mismatch	3' AC CTA TTT G**CA** TGA GAT AC-Cy3 5'

2.3. Characterization Techniques

The SnO$_2$ film morphology was studied using scanning electron microscopy (XL30, Philips, Eindhoven, Netherlands) and transmission electron microscopy (JEOL 2010, Tokyo, Japan). TEM and electron diffraction were carried out at 200 kV with a 0.19-nm point-to point resolution. Cross-section samples were obtained by the tripod method. Samples were polished on both sides using diamond impregnated films. Low-angle ion Ar$^+$ beam milling was used for final perforation of the samples and to minimize contamination.

Impedance measurements were carried out: (I) on the bare electrodes; (II) after silanization step; (III) on the DNA probe grafted electrodes before and (IV) after DNA hybridization. The electrolyte used systematically was the pure hybridization buffer solution, containing no DNA target. A laboratory-made microfluidics cell involving a plexiglas three electrode set-up was used. In this cell, the liquid volume is 500 μL. The circular and functional surface of the film which acts as the working electrode is 0.19 cm^2. The reference electrode is Ag/AgCl (ref.), and the counter-electrode is platinum. The electrodes are connected to a Versatile Simple Potentiostat (VSP,) impedance-analyzer (Bio-Logic, Claix, France). For EIS measurements, this apparatus is used between 10 mHz to 200 kHz with a modulation of 10 mV and an applied voltage of −0.5 V (*vs.* ref.). The impedance spectra were analyzed with Z-fit within the EC-lab software (Bio-Logic, Claix, France) using Non-linear Least Squares Fit principles.

Although this study is ultimately aimed at the development of DNA hybridization techniques which avoid the use of any label, the use of the Cy3 labelled DNA target for the impedance measurements allows the DNA hybridization validation and the systematic comparison of electrical results with the complementary optical results (fluorescence). Epifluorescence measurements were achieved using an BX41M microscope (Olympus, Tokyo, Japan), fitted with a 100 W mercury lamp, a cyanide Cy3 dichroic cube filter (excitation 550 nm, emission 580 nm) and a cooled Spot RT monochrome camera (Diagnostic, Sterling Heights, MI, USA). The Image Pro plus software (Olympus, Tokyo, Japan) was used for image analysis. The fluorescence intensity is measured at two distinct regions of the sample: the spot where DNA probes were grafted and the background outside the spot where no DNA probe was immobilized. This background intensity was then subtracted from the intensity of each spot. The fluorescence intensity value for each condition represents the average over nine different acquisitions from two independent samples.

3. Results and Discussion

3.1. Bare Electrodeposited SnO$_2$ Film Characteristics

The morphology of films electrodeposited with different charge densities, *i.e.*, 0.2, 0.4 and 0.8 C·cm^{-2} is revealed through typical SEM images shown in Figure 1a–c. The film thickness, determined from cross-sectional SEM images, increases linearly with the charge density. The thicknesses are 220 ± 20, 380 ± 20 and 940 ± 50 nm, corresponding to charge densities of 0.2, 0.4, and 0.8 C·cm^{-2}, respectively. The top view images (inset) present a porous surface composed of numerous circular nanoparticles. The particle size does not change significantly going from 5 to 20 nm, when increasing the charge density from 0.2 to 0.8 C·cm^{-2}. Due to the difficulty of observing and measuring efficiently the pore size from SEM images, the morphology of the films is further characterized by TEM observation. As expected, the cross-section bright field HRTEM micrograph reveals much better the local porous structure of the film with highly dispersed SnO$_2$ nanoparticles (Figure 1d). It shows many nanocrystallites with clear lattice fringes corresponding to tetragonal SnO$_2$. The average pore size is approximately 10 nm. Besides, the corresponding selected area electron diffraction (SAED) pattern (inset Figure 1d) exhibits two hollow diffraction rings corresponding to the (110) and (101) of tetragonal SnO$_2$. The hollow rings reveal a quasi-amorphous microstructure of the nanoporous film which was also confirmed by grazing incidence angle XRD.

154

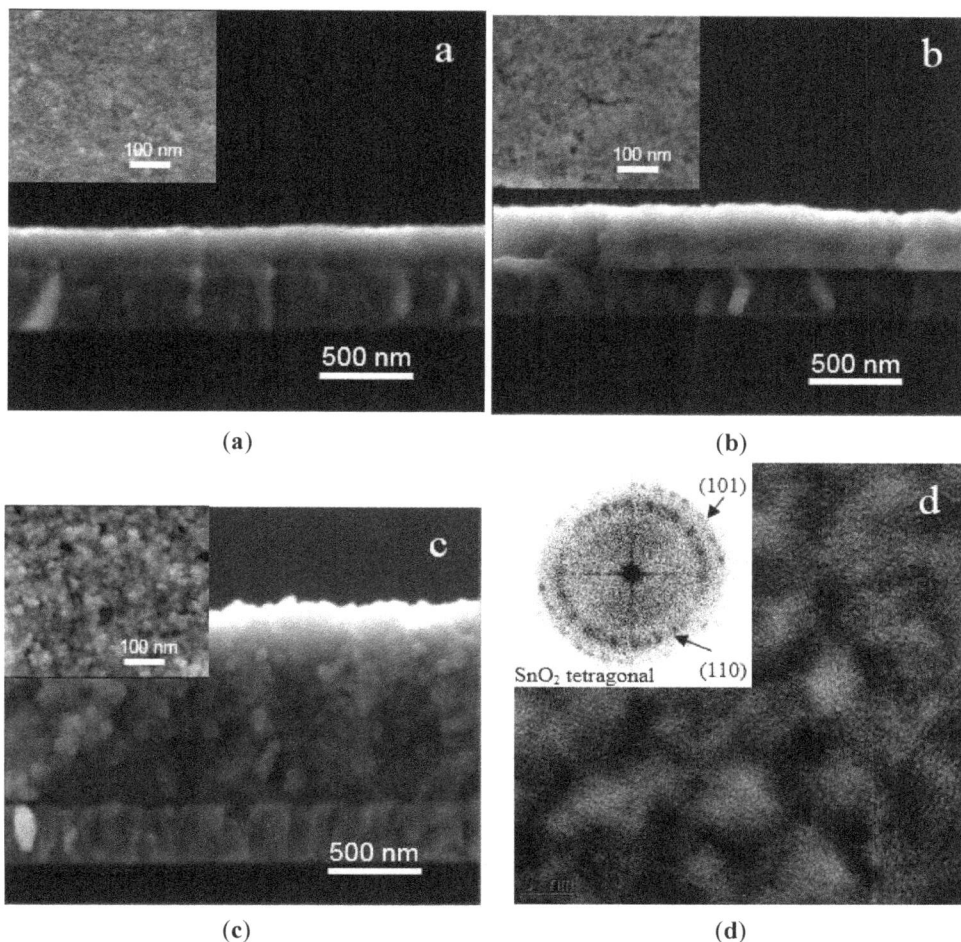

Figure 1. SEM images of SnO_2 films deposited onto ITO substrate at -1.0 V (*vs.* Ag/AgCl) with charge density (Q) of (**a**) 0.2; (**b**) 0.4 and (**c**) 0.8 C·cm^{-2}; (**d**) Typical cross-section HRTEM image of the film deposited with Q of 0.8 C·cm^{-2}. Inset shows the corresponding SAED pattern.

3.2. DNA Hybridization Detection

3.2.1. Influence of Film Thickness on the DNA Hybridization Detection Signals

The validation of DNA hybridization on all nanoporous SnO_2 films was first performed using epifluorescence microscopy. Figure 2a presents a typical top view image after DNA hybridization with Cy3 labelled complementary DNA targets on a 380-nm-thick SnO_2 film. Two observations can be made. First,

155

at the centre of the DNA drop spot, the fluorescence intensity distribution is discontinuous and discrete. Second, the border of the drop is not sharp and a fluorescence intensity gradient is observed. If these observations are similar to the ones obtained on SnO_2 nanopillars [34], they differ from the ones obtained on dense 2D SnO_2 thin film electrode which provided a homogeneous intensity inside the DNA drop with a sharp border [9]. The difference should be associated to the nanostructured morphology which considerably modifies and enhances the hydrophilic characteristic of the surface compared to a 2D thin film surface, causing a capillary effect and a spreading of the DNA droplet on the nanoporous surface. The fluorescence intensity significantly depends on the film thickness. The thicker the film is, the more contrasted the DNA drop is, indicating a higher amount of hybridized DNA. As a result the 940 nm thick film shows the highest fluorescence signal, *i.e.*, 1640 ± 200, while the fluorescence intensities are 270 ± 30 and 1010 ± 120 in the case of 220 nm and 380 nm thick films, respectively. To confirm that the fluorescence signal actually comes from DNA hybridization, the hybridization procedure was also carried with the hybridization solution buffer containing either no DNA target or non-complementary DNA target molecules. In the latter case, the results showed a negligible fluorescence signal coming from non-specific adsorption of DNA target (Figure 2b) while in the former case, no fluorescence signal was detected. The obtained results demonstrate the success and specificity of the used DNA hybridization process on the porous SnO_2 films.

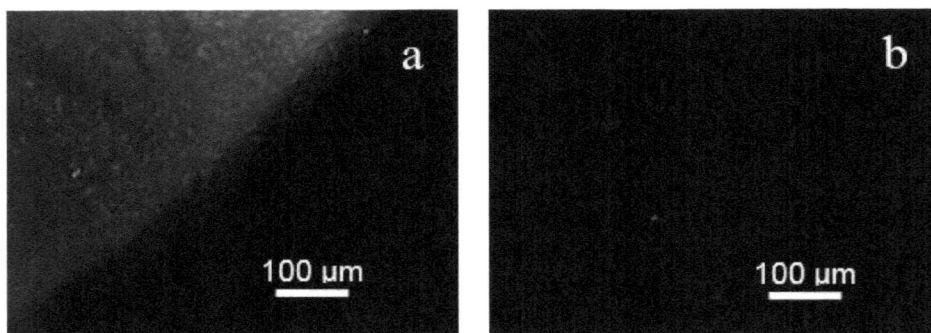

Figure 2. Typical epifluorescence micrographs showing the border of DNA drop on 380-nm-thick SnO_2 nanoporous films after hybridization with Cy3 labelled targets in the case of (**a**) complementary and (**b**) non-complementary DNA hybridization with target concentration of 2 μM.

In a second step, the electrochemical behaviour of nanoporous SnO_2 films with increasing thicknesses was studied (I) on the bare films and after each main step of the functionalization process: (II) after film liquid phase silanization; (III) after probe grafting (ss-DNA) and (IV) after DNA hybridization (ds-DNA). To validate that the impedance variations actually originates from DNA hybridization, impedance analyses were performed in the case of both complementary and non-complementary hybridization.

Whatever the film thickness, the Nyquist plots of bare nanoporous SnO_2 films (Figure 3) display a semi-circular shape. Besides, the semicircle diameter decreases when increasing the film thickness. The impedance of nanoporous SnO_2 film electrode can be analysed by a simple equivalent circuit R_e (R_1, Q_1). The resistance R_e is the sum of ohmic resistances of both the electrolyte bulk and the electrode (ITO with SnO_2 bulk). The parallel element circuit (Q_1, R_1) responsible for the observed semicircle can be essentially attributed to the polarization of the SnO_2/electrolyte interface. Because the obtained semicircles of the Nyquist diagrams present a non-completely symmetric shape, which is due to some non-ideal behaviour, the use of a CPE instead of a capacitor is required. The impedance of a CPE is given by $Z_{CPE} = (j\omega)^{-\alpha}/C$ where α is an empirical coefficient.

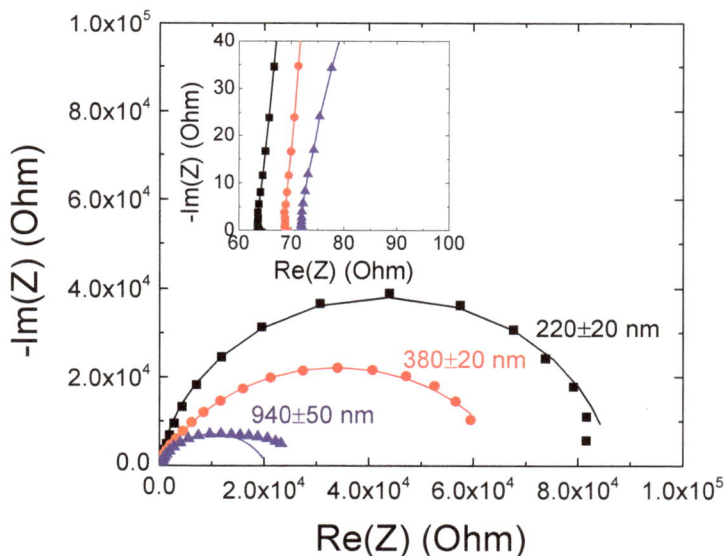

Figure 3. Nyquist plot (recorded at -0.5 V *vs.* ref.) of nanoporous SnO_2/ITO electrodes with increasing film thicknesses from 220 ± 20 to 940 ± 50 nm. The filled symbols correspond to experimental data and the continuous lines to fitting data. The inset shows a zoom-in of the impedance curves at high frequency region.

Extracted electrical parameters from the modelling (Table 2) showed that the resistance R_e increases from 64.6 to 73.9 Ω when the film thickness increases from 220 ± 20 to 940 ± 50 nm. The increase of R_e is mainly due to the increase of the SnO_2

film bulk resistance with the film thickness. However, R_1 decreases sharply when the film thickness increases. The drop of R_1 can be explained in term of a higher real surface area in the case of the thicker films. It is believed that the increment of real surface area improves the ionic interaction at electrolyte-electrode interface resulting in low R_1.

Table 2. Electrical parameter values obtained from fitting of Nyquist plot of bare nanoporous SnO_2 electrodes with increasing film thickness.

Q (C/cm^2)	Film Thickness (nm)	R_e (Ω)	C_1 (μF)	α_1	R_1 (Ω)
0.2	220±20	64.61	23.37	0.779	84,782
0.4	380±20	70.40	26.29	0.772	66,496
0.8	920±50	73.90	38.15	0.810	21,276

The overall electrochemical behaviour of bio-modified films does not change upon the functionalization step since the corresponding Nyquist plots still exhibit one large semicircle. However, their corresponding diameters undergo significantly change upon the modification step as it is shown in the case of complementary DNA hybridization for a 220 nm thick SnO_2 film (Figure 4a) as well as in the case of non-complementary hybridization (Figure 4b). The changes are induced by the different molecular layers immobilized on the film surface. The silanization induces a large increase of the semicircle diameter, while the DNA probe grafting results in a decrease of the semicircle diameter, which is amplified upon the complementary DNA hybridization (Figure 4a). However, this last impedance change is weak in the case of non-complementary hybridization (Figure 4b). These electrochemical behaviours were systematically found for all studied films whatever the film thickness.

As for bare SnO_2 films, all impedance curves are best fitted with an equivalent circuit $R_e(R_1, Q_1)$. In this study, we focused on the evolution of the real part of the impedance, namely the resistance R_1. Its value could be obtained by extrapolating the fit up to the real axis. By monitoring the changes of R_1 we can get information about the different modification steps of the SnO_2 nanoporous based DNA sensors. We have calculated the variation of these resistances expressed as $\Delta R_1/R_1$. $\Delta R_1/R_1 = [R_1$ (after hybridization) $- R_1$ (before hybridization)]$/R_1$ (before hybridization) $\times$ 100%. As expected, R_1 significantly varied upon modification step.

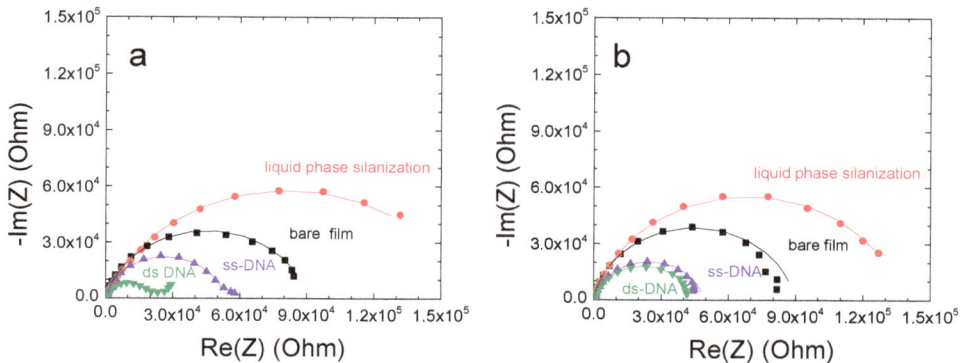

Figure 4. Nyquist plots of bio-modified 220 ± 20 nm thick SnO_2 nanoporous films. DNA hybridization was performed with (**a**) complementary and (**b**) non-complementary DNA targets with target concentration of 2 μM. The filled symbols correspond to experimental data and the continuous lines to fitting data.

Table 3. Resistance value R_1 obtained from fitting of the Nyquist plot of nanoporous SnO_2 electrodes with different film thickness after liquid phase silanization, after immobilization and after complementary or non-complementary DNA hybridization with target concentration of 2 μM.

Film Thickness (nm)	DNA Target Molecules	R_1 (Ω)				$\Delta R_1/R_1$ (%)
		SnO_2 Film	Silanized	ss_DNA	ds_DNA	
220 ± 20	complementary	84782	146577	51974	24743	-54 ± 5
	non-complementary	87250	136078	45541	42543	-6 ± 2
380 ± 20	complementary	66496	115162	95137	43301	-59 ± 5
	non-complementary	67596	127038	101959	95300	-5 ± 2
940 ± 50	complementary	21276	47176	23047	15413	-33 ± 4
	non-complementary	20486	56258	28471	26605	-6 ± 2

The R_1 values from equivalent circuit are reported in Table 3 for both films after different functionalization steps, in the case of complementary and non-complementary DNA hybridization. After silanization, the resistance R_1 increases considerably. It could be due to the coverage of the non-charged and hydrophobic APTES layer on the electrode surface which blocks the electrolyte from diffusing within the porous layer. After ss-DNA grafting onto the silanized surface, the resistance R_1 decreases. The negatively charged ss-DNA presumably trapped inside the nanoporous structure could facilitate the ionic current between the electrolyte and the electrode. Finally, the DNA hybridization with complementary target DNA molecules results in an additional decrease of R_1. The decrease of R_1 upon DNA hybridization could be explained by the observed hydrophilic character and the change of conformation linked to double-stranded ds-DNA. On the one hand, the hydrophilic ds-DNA could partially facilitate some ionic molecules of electrolyte to reach the electrode surface following their infiltration into the nanoporous structure [19]. On the other hand, the conformation of DNA changes from random coil for ss-DNA to a rigid helicoidal chain after hybridization [37,38]. Therefore, it is believed that the electrode surface could be more liberated after hybridization.

The decrease of the polarization resistance R_1 is about $-33\% \pm 4\%$ for the thickest film (940 ± 50 nm) and about $-54\% \pm 5\%$ for the thinnest film (220 ± 20 nm). We note that whatever the film thickness, in the case of non-complementary hybridization, a weak decrease of R_1 is obtained, *i.e.*, about $6\% \pm 2\%$. In the case of complementary hybridization, we observe that if the change of impedance is rather similar for the thin films (220 and 380 nm), it drops for the thickest film (940 nm). We attempt to explain this drop by making a relation with the percentage of surface area which is influenced by DNA hybridization. We hypothesize that due to much larger specific surface area, the amount of DNA probe and target molecules absorbed within the thicker film should be much higher than in the thin one, leading to higher fluorescence signal as mentioned above. However, because of very high specific area for the thickest film, the percentage of the surface area on which DNA probe was grafted is lower than for the thinner ones. Consequently, the DNA free surface is higher and the impedance signal becomes less important when increasing the film thickness. From this result, we deduce that the thinnest films are more relevant for observing impedance changes upon hybridization. For this reason, we follow further experimentations using the 220 nm thick films.

Besides, interesting comparisons can be made with our previous results obtained for 2D SnO_2 dense film electrodes [32] and 1D SnO_2 nanopillar electrodes [34]. It is to be reminded that our research work focuses on the improvement of the sensitivity performances of 2D SnO_2 material by taking the advantage of higher developed surface of SnO_2 nanostructured electrodes. From the results, it is clear that the sensitivity of the 0D-nanoporous film-based DNA sensors compared to that of 2D dense SnO_2 film ($-59\% \pm 5\%$ *vs.* $50\% \pm 10\%$) does not improve as much as that of 1D SnO_2-nanopillar electrode ($97\% \pm 7\%$) for a DNA target concentration of 2.0 μM. However these results emphasize the importance of both the dimensional and morphological organizations of the sensing material on the impedimetric signal upon DNA hybridization. In both cases, two similar behaviours are found. First, the effect of silanization results in a large increase of the impedance, due to the non-charged APTES molecules which block the electrode surface. Second, the DNA probe grafting results in a decrease of the impedance which confirms the presence of charged molecules on both surfaces. However, we observe an opposite behaviour of the impedance upon DNA hybridization. Here it showed a decrease while it showed an increase in the case of 2D and 1D SnO_2 electrodes. To explain this different tendency, it is to be considered that the interfacial charge distribution is different according to the electrode morphology. Regarding the 0D nanoporous films, the DNA strands and the ionic species infiltrated and are trapped within the film thickness, while they are located above the 1D nanopillars and 2D dense film surface. Generally, in the case of non-Faradic detection, DNA hybridization can induce a change of the impedance in several manners in relation with either intrinsic or external causes.

On the one hand, in the case of 1D nanopillars and 2D dense films, the increase of the impedance upon DNA hybridization can be explained by a cause which is intrinsic to the SnO_2 material, namely, the field effect. The addition of negatively charged DNA molecules upon hybridization leads to an increase of the space charge thickness which is located below the film surface (in the case of 2D dense film) and below the nanopillar surface. On the other hand, in the case of 0D nanoporous films, the decrease of the impedance upon DNA hybridization can be explained by some external phenomena as discussed above. The penetration of hydrophilic and charged double-stranded DNA molecules within the nanoporous film volume enhances the transport of ionic species inside the electrode volume. As a result, the impedance of this complex interface is reduced. In this case, the field effect is hindered and does not play any predominant role.

3.2.2. APTES Vapour Phase Deposition *vs.* Liquid Phase Deposition

In order to obtain higher performance DNA sensor, the immobilization of the DNA probes on the film electrode needs to be well controlled. In our work, the DNA probes are covalently grafted to the aminosilane (APTES) through a cross-linker (glutaraldehyde). Functionalized surfaces were created by chemical treatment using silanization process which was first carried out in our laboratory by liquid phase deposition of a solution of silane diluted in 95% pure ethanol and 5% deionized water. However, the main issue of liquid treatment is the eventual ability of the precursor to copolymerize in the presence of water forming an inhomogeneous organosilane monolayer on the surface [39]. To overcome this problem, the vapour phase deposition has been performed in a next step. The low density of the agent in vapour phase could reduce the aggregation formation. Importantly, because the vapour is more effective than the solution in penetrating into the nanoporous structure of the films, it is expected that a superior organosilane monolayer is achieved and consequently, a better DNA surface coverage. As a result, the DNA detection performance should be enhanced.

The Nyquist plots obtained on 220 ± 20 nm thick SnO_2 films in the case of vapour phase silanization (Figure 5a) clearly show the importance of the silanization conditions when comparing with liquid phase (Figure 5b). In this case, the semicircle (red curve) presents a much larger diameter than that of liquid phase deposition. As previously, we perform the Nyquist plot modeling by using the equivalent circuit

$R_e(R_1, Q_1)$ to determine the polarization resistance R_1 variation upon the stepwise modification. The resistance R_1 obtained from the impedance curve after vapour phase silanization revealed an approximately three times higher value than the one of the liquid phase silanization (432,768 Ω *vs.* 146,577 Ω in the case of liquid phase deposition). It indicates that the deposited organosilane monolayer from vapour phase was more efficient on the nanoporous film than in the case of liquid phase. DNA hybridization is then performed on both film surfaces with the same DNA target concentration of 2 μM. The change of resistance $\Delta R_1/R_1$ upon DNA hybridization increases from $-54\% \pm 5\%$ in the case of liquid phase deposition (Table 3) to $-63\% \pm 5\%$ of vapour phase deposition (Table 4).

Table 4. Vapour phase and liquid phase silanization: resistance value R_1 obtained from fitting experimental data to the equivalent circuit for 220-nm-thick-nanoporous-SnO$_2$ electrodes after silanization, after DNA probe immobilization and after complementary DNA hybridization with different target concentrations.

	Vapour Phase Silanization				
$C_{DNA\ target}$ (μM)	R_1 (Ω)				$\Delta R_1/R_1$ (%)
	SnO$_2$ Film	Silanized	ss_DNA	ds_DNA	
2.0	96,782	43,2768	167,202	61,836	-63 ± 5
1.0	86,453	370,742	147,929	76,689	-48 ± 5
0.5	85,310	377,590	143,687	97,006	-33 ± 3
0.1	81,101	416,567	121,467	99,531	-18 ± 3
0.01	79,987	393,879	126,847	111,929	-11 ± 3
	Liquid Phase Silanization				
$C_{DNA\ target}$ (μM)	R_1 (Ω)				$\Delta R_1/R_1$ (%)
	SnO$_2$ Film	Silanized	ss_DNA	ds_DNA	
1.0	97,419	144,626	47,492	34,015	-28 ± 5
0.5	97,926	151,935	45,305	37,707	-17 ± 3
0.1	95,509	135,238	43,923	40,441	-7 ± 2
0.01	95,264	141,849	48,337	47,538	-2 ± 1

This nearly 10% increase of the EIS signal confirms that the sensitivity of the DNA detection could be improved significantly by using vapour phase silanization process. The EIS result was confirmed by fluorescence measurements carried out on the corresponding samples. It showed almost three times higher fluorescence intensity in the case of vapour phase deposition over the liquid method (insets of Figure 5).

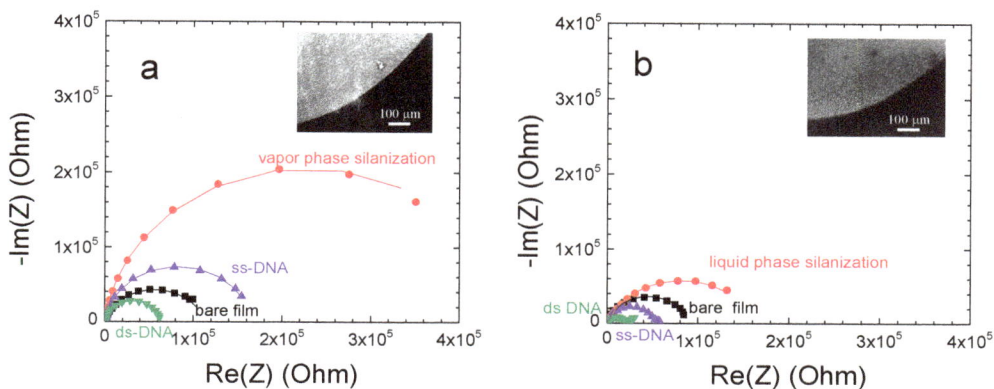

Figure 5. Nyquist plots of 220 nm thick bio-modified SnO_2 nanoporous film with (**a**) vapour phase and (**b**) liquid phase silanization processes. DNA complementary hybridization was performed with target concentration of 2 µM. The filled symbols correspond to experimental data and the continuous lines to fitting data. The inset shows the corresponding typical fluorescence micrograph after hybridization of complementary target labelled with Cy3.

The sensitivity of the biomodified 220 nm thick SnO_2 films was studied by detecting complementary DNA target at lower concentrations. The evolution of the polarization resistance ratio $\Delta R_1/R_1$ has been plotted as a function of DNA target concentration (Figure 6) for both silanization processes: vapour and liquid phase deposition. The lower DNA target concentration, the less important is the decrease of resistance R_1. In the case of vapour phase silanization, the decrease of the polarization resistance $\Delta R_1/R_1$ is systematically wider, from 10% to 20%, than in the case of liquid phase silanization. Indeed, in the first case it ranges from $-48\% \pm 5\%$ to $-11\% \pm 3\%$ when decreasing DNA target concentration from 1 µm down to 10 nM (Table 4), whereas, it decreases only from $-28\% \pm 5\%$ to $-2\% \pm 1\%$ in the case of liquid phase silanization (Table 4). As expected, even for low DNA target concentrations, the infiltration of organosilane molecules into the nanopores is facilitated in the case of gas phase, which plays a role in the sensitivity enhancement of the DNA sensor.

Finally the selectivity of the sensor on the 220 nm thick nanoporous SnO_2 based DNA sensor has further been tested by performing hybridization procedure with 1- and 2-base-mismatch DNA target molecules as well as with blank hybridization (buffer with no DNA target molecule). The concentration of all DNA targets was fixed at 2 µM. The silanization was carried out only in vapour phase deposition technique. Impedance curves exhibit one semicircles for all kinds of target molecule. The impedance curves were analysed in terms of R_1 variations. As expected, $\Delta R_1/R_1$ varies differently following the types of DNA target as can be seen from Table 5 and in Figure 7. $\Delta R_1/R_1$ was equal to $-34\% \pm 5\%$ and $-21\% \pm 4\%$ in the case of 1- and 2-base-mismatch DNA targets, respectively. A negligible signal ($-1\% \pm 0.5\%$) is obtained upon a blank hybridization. Such a value can be considered as background signal.

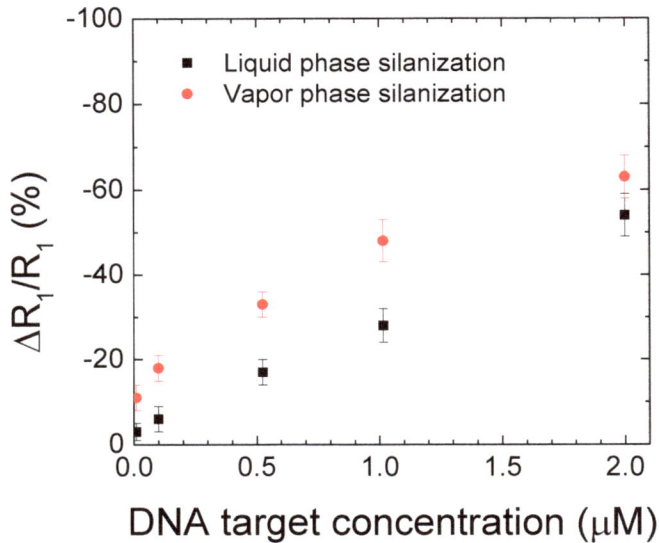

Figure 6. Evolution of the polarization resistance $\Delta R_1/R_1$ of the biomodified SnO_2 nanoporous films as a function of target DNA concentration in the case of vapour (red) and liquid phase silanization (black).

To further support the results observed in impedance measurements, the selectivity of the films has also been studied optically by epifluorescence optical microscopy. The evolutions of both fluorescence intensity and variation of resistance $\Delta R_1/R_1$ respectively of the bio-modified 220 nm thick nanoporous-SnO_2 electrode as a function of different types of DNA target are shown in Figure 7. The complementary hybridization gives the highest fluorescence signal, *i.e.*, 920 ± 70, while the fluorescence intensity significantly drops when 1- and 2-base mismatch DNA targets are used, *i.e.*, 180 ± 20 and 100 ± 20, respectively. Non-complementary hybridization provides a negligible signal, *i.e.*, 10 ± 5 of non-specific adsorption of DNA target. In the case of blank hybridization, the area where the DNA probes are immobilized could not be found. The fluorescence results matched rather well with those of impedance, which demonstrates the high selectivity of the process on nanoporous SnO_2 sensing matrix.

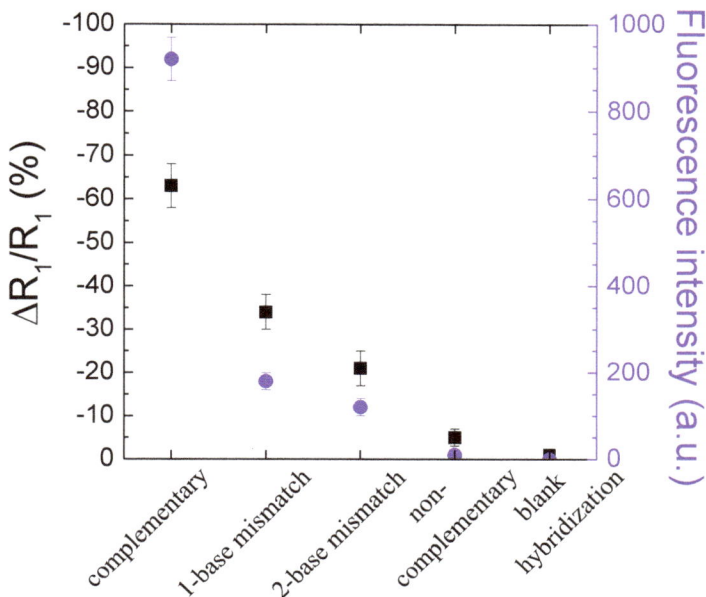

Figure 7. Evolutions of the polarization resistance $\Delta R_1/R_1$ (%) (black) and the fluorescence signal (blue) as a function of different types of DNA target: complementary, 1-, 2-base mismatch, non-complementary and hybridization buffer without DNA target (blank hybridization).

Table 5. Vapour phase silanization: resistance value R_1 obtained from fitting experimental data to the equivalent circuit for DNA probe immobilized 220-nm-thick-nanoporous-SnO_2 electrodes after silanization, after DNA probe immobilization and after complementary DNA hybridization with different types of target molecules. The target concentration was fixed at 2 µM.

DNA Target Molecule	$C_{DNA\ target}$ (µM)	R_1 (Ω)			$\Delta R_1/R_1$ (%)
		SnO_2 Film	Silanized ssDNA	dsDNA	
1-base mismatch		98742	356254	120292	79007	-34 ± 5
2-base mismatch	2.0	97249	374519	109624	86879	-21 ± 4
Blank hybridization (buffer without DNA target)		87876	381906	105884	99632	-1 ± 0.5

4. Conclusions

We have studied the label free DNA detection using EIS on 0D nanoporous SnO_2 films that have been deposited by an electrodeposition process. The films thickness has been varied from 220 ± 20 to 940 ± 50 nm. The results have shown a systematic decrease of the impedance upon DNA hybridization, the decrease being more pronounced for the thinnest films. The decrease of the impedance upon DNA hybridization has been attributed to the enhanced penetration of ionic species within the film volume.

The comparison of impedance variations upon DNA hybridization between the liquid and vapour phase processes for APTES grafting on the nanoporous SnO_2 films showed that vapour-phase method is more efficient. This is due to the fact that the vapour is more effective than the solution in penetrating into the films' nanopores. As a result, the DNA sensors made with a vapour-treated silane layer exhibit a higher sensitivity than those produced from liquid-treated silane, in the range of tested target DNA concentrations, going to 10 nM. Finally, the impedance and fluorescence response signals strongly depend on the types of target DNA molecules, demonstrating a high selectivity of the process on nanoporous SnO_2 films.

Acknowledgments: The authors express our thanks to Anh Tuan Mai from ITIMS (HUST, Vietnam) for helpful discussion, Didier Delabouglise from LMGP (UGA) for his useful help in the electrodeposition of the SnO2 porous films on ITO/glass substrate, Eng. Beatrice Doisneau and Eng. Laetitia Rapenne for their helps in SEM and TEM observations.

Author Contributions: All authors collaborated to carry out the work presented here. Valerie Stambouli defined the research topic. Minh Hai Le carried out the experiments. Valerie Stambouli and Minh Hai Le interpreted the results and wrote the paper. Eric Chainet and Carmen Jimenez helped prepare the electrodeposition system, reviewed and edited the manuscript. All authors read and approved the manuscript.

Conflicts of Interest: The authors declare no conflict of interest.

References

1. Sassolas, A.; Leca-Bouvier, B.D.; Blum, L.J. DNA Biosensors and Microarrays. *Chem. Rev.* **2008**, *108*, 109–139.
2. Livache, T.; Bazin, H.; Mathis, G. Conducting polymers on microelectronic devices as tools for biological analyses. *Clin. Chim. Acta* **1998**, *278*, 171–176.
3. Kushon, S.A.; Ley, K.D.; Bradford, K.; Jones, R.M.; McBranch, D.; Whitten, D. Detection of DNA hybridization via fluorescent polymer superquenching. *Langmuir* **2002**, *18*, 7245–7249.
4. Azek, F.; Grossiord, C.; Joannes, M.; Limoges, B.; Brossier, P. Hybridization Assay at a Disposable Electrochemical Biosensor for the Attomole Detection of Amplified Human Cytomegalovirus DNA. *Anal. Biochem.* **2000**, *284*, 107–113.
5. Fortin, E.; Mailley, P.; Lacroix, L.; Szunerits, S. Imaging of DNA hybridization on microscopic polypyrrole patterns using scanning electrochemical microscopy (SECM): The HRP bio-catalyzed oxidation of 4-chloro-1-naphthol. *Analyst* **2006**, *131*, 186–193.
6. Ferreira, H.A.; Cardoso, F.A.; Ferreira, R.; Cardoso, S.; Freitas, P.P. Magnetoresistive DNA chips based on ac field focusing of magnetic labels. *J. Appl. Phys.* **2006**, *99*.
7. Cai, H.; Wang, Y.; He, P.; Fang, Y. Electrochemical detection of DNA hybridization based on silver-enhanced gold nanoparticle label. *Anal. Chim. Acta* **2002**, *469*, 165–172.

8. Peng, H.; Soeller, C.; Cannell, M.B.; Bowmaker, G.A.; Cooney, R.P.; Travas-Sejdic, J. Electrochemical detection of DNA hybridization amplified by nanoparticles. *Biosens. Bioelectron.* **2006**, *21*, 1727–1736.

9. Stambouli, V.; Labeau, M.; Matko, I.; Chenevier, B.; Renault, O.; Guiducci, C.; Chaudouët, P.; Roussel, H.; Nibkin, D.; Dupuis, E. Development and functionalisation of Sb doped SnO_2 thin films for DNA biochip applications. *Sens. Actuators B Chem.* **2006**, *113*, 1025–1033.

10. Daniels, J.S.; Pourmand, N. Label-Free Impedance Biosensors: Opportunities and Challenges. *Electroanalysis* **2007**, *19*, 1239–1257.

11. Ansari, A.A.; Singh, R.; Sumana, G.; Malhotra, B.D. Sol–gel derived nano-structured zinc oxide film for sexually transmitted disease sensor. *Analyst* **2009**, *134*, 997.

12. Patel, M.K.; Singh, J.; Singh, M.K.; Agrawal, V.V.; Ansari, S.G.; Malhotra, B.D. Tin Oxide Quantum Dot Based DNA Sensor for Pathogen Detection. *J. Nanosci. Nanotechnol.* **2013**, *13*, 1671–1678.

13. Popovich, N.D.; Eckhardt, A.E.; Mikulecky, J.C.; Napier, M.E.; Thomas, R.S. Electrochemical sensor for detection of unmodified nucleic acids. *Talanta* **2002**, *56*, 821–828.

14. Yang, I.V.; Thorp, H.H. Modification of Indium Tin Oxide Electrodes with Repeat Polynucleotides: Electrochemical Detection of Trinucleotide Repeat Expansion. *Anal. Chem.* **2001**, *73*, 5316–5322.

15. Park, J.Y.; Park, S.M. DNA hybridization sensors based on electrochemical impedance spectroscopy as a detection tool. *Sensors* **2009**, *9*, 9513–9532.

16. Lazerges, M.; Bedioui, F. Analysis of the evolution of the detection limits of electrochemical DNA biosensors. *Anal. Bioanal. Chem.* **2013**, *405*, 3705–3714.

17. Christine Berggren, P.S. A Feasibility Study of a Capacitive Biosensor for Direct Detection of DNA Hybridization. *Electroanalysis* **1999**, *11*, 156–160.

18. Lai, W.A.; Lin, C.H.; Yang, Y.S.; Lu, M.S.C. Ultrasensitive and label-free detection of pathogenic avian influenza DNA by using CMOS impedimetric sensors. *Biosens. Bioelectron.* **2012**, *35*, 456–460.

19. Tlili, C.; Korri-Youssoufi, H.; Ponsonnet, L.; Martelet, C.; Jaffrezic-Renault, N.J. Electrochemical impedance probing of DNA hybridisation on oligonucleotide-functionalised polypyrrole. *Talanta* **2005**, *68*, 131–137.

20. Gautier, C.; Cougnon, C.; Pilard, J.F.; Casse, N. Label-free detection of DNA hybridization based on EIS investigation of conducting properties of functionalized polythiophene matrix. *J. Electroanal. Chem.* **2006**, *587*, 276–283.

21. Gautier, C.; Esnault, C.; Cougnon, C.; Pilard, J.F.; Casse, N.; Chénais, B. Hybridization-induced interfacial changes detected by non-Faradaic impedimetric measurements compared to Faradaic approach. *J. Electroanal. Chem.* **2007**, *610*, 227–233.

22. Souteyrand, E.; Cloarec, J.P.; Martin, J.R.; Wilson, C.; Lawrence, I.; Mikkelsen, S.; Lawrence, M.F. Direct Detection of the Hybridization of Synthetic Homo-Oligomer DNA Sequences by Field Effect. *J. Phys. Chem. B* **1997**, *101*, 2980–2985.

23. Cloarec, J.P.; Martin, J.R.; Polychronakos, C.; Lawrence, I.; Lawrence, M.F.; Souteyrand, E. Functionalization of Si/SiO$_2$ substrates with homooligonucleotides for a DNA biosensor. *Sens. Actuators B Chem.* **1999**, *58*, 394–398.

24. Chen, C.P.; Ganguly, A.; Lu, C.Y.; Chen, T.Y.; Kuo, C.C.; Chen, R.S.; Tu, W.H.; Fischer, W.B.; Chen, K.H.; Chen, L.C. Ultrasensitive *in situ* Label-Free DNA Detection Using a GaN Nanowire-Based Extended-Gate Field-Effect-Transistor Sensor. *Anal. Chem.* **2011**, *83*, 1938–1943.

25. Cai, W.; Peck, J.R.; van der Weide, D.W.; Hamers, R.J. Direct electrical detection of hybridization at DNA-modified silicon surfaces. *Biosens. Bioelectron.* **2004**, *19*, 1013–1019.

26. Vamvakaki, V.; Chaniotakis, N.A. DNA Stabilization and Hybridization Detection on Porous Silicon Surface by EIS and Total Reflection FT-IR Spectroscopy. *Electroanalysis* **2008**, *20*, 1845–1850.

27. Yang, W.; Butler, J.E.; Russell, John N.; Hamers, R.J. Interfacial Electrical Properties of DNA-Modified Diamond Thin Films: Intrinsic Response and Hybridization-Induced Field Effects. *Langmuir* **2004**, *20*, 6778–6787.

28. Vermeeren, V.; Bijnens, N.; Wenmackers, S.; Daenen, M.; Haenen, K.; Williams, O.A.; Ameloot, M.; vandeVen, M.; Wagner, P.; Michiels, L. Towards a Real-Time, Label-Free, Diamond-Based DNA Sensor. *Langmuir* **2007**, *23*, 13193–13202.

29. Chen, C.P.; Ganguly, A.; Wang, C.H.; Hsu, C.W.; Chattopadhyay, S.; Hsu, Y.K.; Chang, Y.C.; Chen, K.H.; Chen, L.C. Label-Free Dual Sensing of DNA Molecules Using GaN Nanowires. *Anal. Chem.* **2009**, *81*, 36–42.

30. Zebda, A.; Stambouli, V.; Labeau, M.; Guiducci, C.; Diard, J.P.; le Gorrec, B. Metallic oxide CdIn$_2$O$_4$ films for the label free electrochemical detection of DNA hybridization. *Biosens. Bioelectron.* **2006**, *22*, 178–184.

31. Zebda, A.; Labeau, M.; Diard, J.P.; Lavalley, V.; Stambouli, V. Electrical resistivity dependence of semi-conductive oxide electrode on the label-free electrochemical detection of DNA. *Sens. Actuators B Chem.* **2010**, *144*, 176–182.

32. Stambouli, V.; Zebda, A.; Appert, E.; Guiducci, C.; Labeau, M.; Diard, J.-P.; Le Gorrec, B.; Brack, N.; Pigram, P.J. Semiconductor oxide based electrodes for the label-free electrical detection of DNA hybridization: Comparison between Sb doped SnO$_2$ and CdIn$_2$O$_4$. *Electrochimica Acta* **2006**, *51*, 5206–5214.

33. Abdelkader, Z. Propriétés Microstructurales et Electriques D'Electrodes D'Oxydes SnO$_2$ et CdIn$_2$O$_4$: Application à la Détection Electrochimique Directe de L'Hybridation de l'ADN. Ph.D. Thesis, INPG, Grenoble, France, 2007.

34. Le, M.H.; Fradetal, L.; Delabouglise, D.; Mai, A.T.; Stambouli, V. Fluorescence and label free impedimetric DNA detection on SnO$_2$ nanopillars. *Electroanalysis* **2015**. in press.

35. Therese, G.H.A.; Kamath, P.V. Electrochemical Synthesis of Metal Oxides and Hydroxides. *Chem. Mater.* **2000**, *12*, 1195–1204.

36. Chang, S.T.; Leu, I.C.; Hon, M.H. Preparation and Characterization of Nanostructured Tin Oxide Films by Electrochemical Deposition. *Electrochem. Solid -State Lett.* **2002**, *5*, C71–C74.

37. Piro, B.; Haccoun, J.; Pham, M.C.; Tran, L.D.; Rubin, A.; Perrot, H.; Gabrielli, C. Study of the DNA hybridization transduction behavior of a quinone-containing electroactive polymer by cyclic voltammetry and electrochemical impedance spectroscopy. *J. Electroanal. Chem.* **2005**, *577*, 155–165.

38. Anne, A.; Bouchardon, A.; Moiroux, J. 3'-ferrocene-labeled oligonucleotide chains end-tethered to gold electrode surfaces: Novel model systems for exploring flexibility of short DNA using cyclic voltammetry. *J. Am. Chem. Soc.* **2003**, *125*, 1112–1113.

39. Fiorilli, S.; Rivolo, P.; Descrovi, E.; Ricciardi, C.; Pasquardini, L.; Lunelli, L.; Vanzetti, L.; Pederzolli, C.; Onida, B.; Garrone, E. Vapor-phase self-assembled monolayers of aminosilane on plasma-activated silicon substrates. *J. Colloid Interface Sci.* **2008**, *321*, 235–241.

169

A Sensitive and Selective Label-Free Electrochemical DNA Biosensor for the Detection of Specific Dengue Virus Serotype 3 Sequences

Natália Oliveira, Elaine Souza, Danielly Ferreira, Deborah Zanforlin,
Wessulla Bezerra, Maria Amélia Borba, Mariana Arruda, Kennya Lopes,
Gustavo Nascimento, Danyelly Martins, Marli Cordeiro and José Lima-Filho

Abstract: Dengue fever is the most prevalent vector-borne disease in the world, with nearly 100 million people infected every year. Early diagnosis and identification of the pathogen are crucial steps for the treatment and for prevention of the disease, mainly in areas where the co-circulation of different serotypes is common, increasing the outcome of dengue hemorrhagic fever (DHF) and dengue shock syndrome (DSS). Due to the lack of fast and inexpensive methods available for the identification of dengue serotypes, herein we report the development of an electrochemical DNA biosensor for the detection of sequences of dengue virus serotype 3 (DENV-3). DENV-3 probe was designed using bioinformatics software and differential pulse voltammetry (DPV) was used for electrochemical analysis. The results showed that a 22-m sequence was the best DNA probe for the identification of DENV-3. The optimum concentration of the DNA probe immobilized onto the electrode surface is 500 nM and a low detection limit of the system (3.09 nM). Moreover, this system allows selective detection of DENV-3 sequences in buffer and human serum solutions. Therefore, the application of DNA biosensors for diagnostics at the molecular level may contribute to future advances in the implementation of specific, effective and rapid detection methods for the diagnosis dengue viruses.

Reprinted from *Sensors*. Cite as: Oliveira, N.; Souza, E.; Ferreira, D.; Zanforlin, D.; Bezerra, W.; Borba, M.A.; Arruda, M.; Lopes, K.; Nascimento, G.; Martins, D.; Cordeiro, M.; Lima-Filho, J. A Sensitive and Selective Label-Free Electrochemical DNA Biosensor for the Detection of Specific Dengue Virus Serotype 3 Sequences. *Sensors* **2015**, *15*, 15562–15577.

1. Introduction

Dengue fever is the most prevalent vector-borne disease in the world. The World Health Organization (WHO) estimates that some 100 million people are infected every year; however, some studies have predicted that this number could be greatly underestimated, and is actually closer to 390 million [1–3]. The distribution of the disease is mainly in tropical and subtropical regions and recently, it is has been increasingly seen in urban and semi-urban areas. All these factors have contributed to reveal dengue fever as a major international public health problem [1,2,4,5].

The infection is caused by a single stranded RNA-virus (DENV) of about 10.7 kb, which belongs to the *Flaviviridae* family, with approximately four antigenically distinct serotypes (DENV-1–DENV-4) [6,7]. The disease exhibits a wide range of symptoms, such as fever, headache and myalgia, which are the most common in classic dengue. Nevertheless, it can also shows more severe manifestations, like in dengue hemorrhagic fever (DHF) or dengue shock syndrome (DSS), which present life-threatening symptoms, such as bleeding, thrombocytopenia and vascular leakage [8–10].

Early diagnosis and identification of the pathogen are necessary for the prevention and treatment of patients, as well as for the avoidance of new outbreaks and emergence of severe cases of dengue, since it is known that the co-circulation of different serotypes in an area increases the possibility of DHF and DSS outcomes [11,12].

Methods to confirm dengue virus infection may involve detection of the virion, viral RNA, antigens or antibodies [13]. Virus detection by cell culture, viral nucleic acid or antigen detection (nonstructural protein 1 or NS1 antigen) can be used to confirm dengue infection in the acute phase of the illness (0–7 days following the onset of the symptoms) [14,15]. In the later phase of the disease, serologic tests are more applied and preferred for diagnosis, as the sensitivity of virus isolation and antigen reactivity decreases [16]. Viral antigen (NS1) detection assays are rapid, reliable and easy to perform, however, they cannot allow to distinguish between different viral serotypes [17,18].

Viral isolation, although considered the gold standard diagnostic method, is time-consuming and highly complex compared with other direct virus detection techniques [1,19]. On the other hand, the RT-PCR assay is widely used, it allows the detection of low copies of viral genes in less than 48 h [20]. However, both techniques are costly and labor-intensive, but they are more specific than serologic methods used for antibody detection and allow one to differentiate between the various dengue virus serotypes [21].

Application of DNA biosensors has emerged as an alternative method to the current molecular biology techniques [22,23]. These devices consist of a single-stranded DNA molecule (ssDNA) attached to a transducing surface that is able to detect a specific nucleic acid sequence, based on DNA hybridization events. Currently, there is a growing interest in developing label-free methods for DNA detection, considering their rapidness, easiness, low cost and minimal sample preparation requirements, compared to labeling methods, where the properties of the modified macromolecules often change, which may result in total loss of bioactivity or stability [24,25]. Label-free approaches rely on the direct detection of intrinsic electrochemical properties of DNA (e.g., oxidation of purine bases, particularly guanine) or on changes in some of the interfacial properties after hybridization. In addition, interference with the biological recognition between DNA molecules is minimized. Nevertheless, in labeling methods, these undesirable effects are more likely to occur due to steric hindrance and blocking of the binding sites [26–28].

Consequently, since biosensors allow to detect and identify DNA sequences in a fast and simple way, herein we report the first step to develop a cost-effective, sensitive and label-free electrochemical DNA biosensor for the detection of DENV-3 sequences in biological samples, as a part of an ongoing research previously published [29].

2. Experimental Section

2.1. Design of a Specific DENV-3 DNA Probe

The complete genomes of dengue virus serotype 3, corresponding to GenBank accession numbers AY099336, AY099337, AY099338S1, AY099338S2, AY099339S1, AY099339S2, AY099340S1, AY099340S2, AY09934S1, AY099342S1 were obtained from the National Center for Biotechnology Information (NCBI) database. These sequences correspond to strains that were introduced in the American continent, and caused the disease outbreaks in 2002 [30,31]. CLC Main Workbench v.6.0 software was used to analyze common sequences among those dengue genomes, by using an alignment tool. Then, a specific DNA probe for DENV-3 was selected by comparison of the homologous sequences with other organisms, using Basic Local Alignment Search Tool (BLAST). DENV-3 complementary (target) and non-complementary sequences were also designed using the same method.

2.2. Reagents and Materials

All chemicals were of reagent grade quality and were used directly as received without further purification. Tris base was obtained from Promega (Fitchburg, WI, USA) and sodium acetate was obtained from Sigma-Aldrich (St. Louis, MO, USA) DENV-3 probes were purchased as lyophilized powder from IDT Technologies (Coraville, IA, USA). The stock and diluted solutions (25 nM) were prepared in 0.5 M acetate buffer (pH 5.0) and kept frozen. Ultrapure RNAse/DNAse-free water was used in all buffer solutions. After bioinformatics analysis, the following DNA sequences were used in this study:

DENV-3 probe: 5′-TAA CAT CAT CAT GAG ACA GAG C-3′
DENV-3 target: 5′-GCT CTG TCT CAT GAT GAT GTT A-3′
Non-complementary sequence: 5′-TCT CTT GTT TAA GAC AAC AGA G-3′

Human serum used in this study was obtained from blood samples provided by the pathogenic virus collection of Centro de Pesquisas Aggeu Magalhães (CPqAM).

Serum solutions were prepared by centrifugation at 3500 rpm for 5 min at 20 °C (3500 rpm for 5 min), in which the obtained supernatant was collected from each sample, and stored at 23 °C until further used for experiments testing.

2.3. Apparatus

Experiments were carried out using a PGSTAT302 potentiostat (METROHM Autolab, Utrecht, The Netherlands) with the GPES 4.9.007 software as a graphic interface. The electrochemical device was composed by a two-electrode system: A pencil graphite electrode (PGE) as a working electrode and silver/chloride silver electrode as a reference electrode. Each measurement consisted of a cycle of activation/immobilization/hybridization/detection by using a fresh PGE surface. All the experiments were performed in triplicate, at room temperature (23 °C).

2.4. Procedure

2.4.1. Preparation of Electrodes and Pre-Treatment of PGE

PGEs were obtained from Mercur (Santa Cruz do Sul, Brazil), as a pencil graphite lead type 4 B. Briefly, PGEs were produced by cutting graphite lead in pieces of 3 cm and polishing them with an emery polishing disc (Dremel, Mount Prospect, IL,USA). The PGEs were then washed with ultrapure water to remove any contaminant present on the surface of the working electrode. The reference electrode was made by immersing a golden pin into an Ag/AgCl ink (Henkel Acheson, Hemel Hempstead, UK) and dried at 40 °C overnight. The polished surface of PGEs was pre-treated by applying a potential of +1.80 V for 5 min in 0.5 M acetate buffer solution (pH 5.0) [32–34].

2.4.2. DNA Probe Immobilization onto PGE Surface

Immobilization of DENV-3 probe onto the PGE surface was performed by immersing the activated PGE in acetate buffer solution, with different concentrations of DENV-3 probes (250–1000 nM), by applying a fixed potential of +0.5 V for 300 s onto the electrode surface.

2.4.3. DNA Hybridization with Complementary and
Non-Complementary Sequences

The hybridization of the immobilized DNA probe on the electrode (PGE/DENV-3 probe) was performed by immersing the electrode in an Eppendorf tube containing 70 µL of DENV-3 target sequences diluted in acetate buffer. The hybridization reaction was then carried out in a thermomixer, stirring at 300 rpm, under a specific annealing temperature of 52 °C, for 10 min. This same procedure was adopted to evaluate the hybridization of the PGE/DENV-3 probe with non-complementary sequences, as well as buffer solutions containing a mix of both 75 nM of DENV-3 complementary and non-complementary sequences (mixed DNA solution).

2.4.4. Detection of Complementary and Non-Complementary Sequences in Human Serum

As a way to evaluate the efficiency of the system to detect DENV-3 sequences in biological samples on the electrode surface, the complementary and non-complementary DNA sequences were diluted in human serum (75 nM concentration) and the hybridization assay was conducted using the same conditions described previously. This procedure was also adopted for tests human serum solutions mixed with both target and non-complementary sequences.

2.5. Electrochemical Analysis

Differential pulse voltammetry (DPV) was used for electrochemical analysis in this study. Current peaks were recorded after applying a potential range of +0.5 up to +1.2 V at a scan rate of 0.05 V/s onto the electrode surface, which was immersed in 20 mM Tris-HCl buffer (pH 7.0). The raw data obtained with DPV technique was treated using the Savitzky and Golay filter (level 2) of the GPES software, followed by moving the average baseline correction using a peak width of 0.01 V [35].

2.6. Statistical Data Analysis

Experimental data were analyzed with Statistica 8.0 software (StatSoft, Tulsa, OK, USA) using parametric tests; Tukey's test was used to compare multi-independent group data, and a level of $p < 0.05$ was considered significant. The reproducibility of the system was expressed as the coefficient of variation inter-assay (CV), which was calculated over three independent assays on the probe-modified PGEs.

3. Results and Discussion

3.1. Bioinformatics Analysis of DENV-3 DNA Probes

The design of DNA probes is one of the crucial steps in the development of a biosensor, because it determines the specificity of the device [36]. For genosensors, this can be achieved using bioinformatics analysis based on whole genome sequencing, in a way to predict the most specific region that is able to produce a steady double-strand DNA with the pathogen [37,38]. In this work, DNA probes specific for DENV-3 were designed mainly by using CLC Main Workbench software, based on a sequence alignment tool to identify regions of similarity between the dengue strains. After that, DNA sequences from the strains that showed specificity only for DENV-3 were compared with other organism genomes using BLAST tool, in order to exclude any correlations. Finally, the Oligonucleotide Properties Calculator (Oligo Calc) software (Northwestern University, IL, USA) was used to provide physical properties information of the selected DENV-3 sequences, in a way to establish the best match of DNA probe for biosensors. Figure 1 shows a flowchart containing the criteria of selection of DENV-3 probes used in this study.

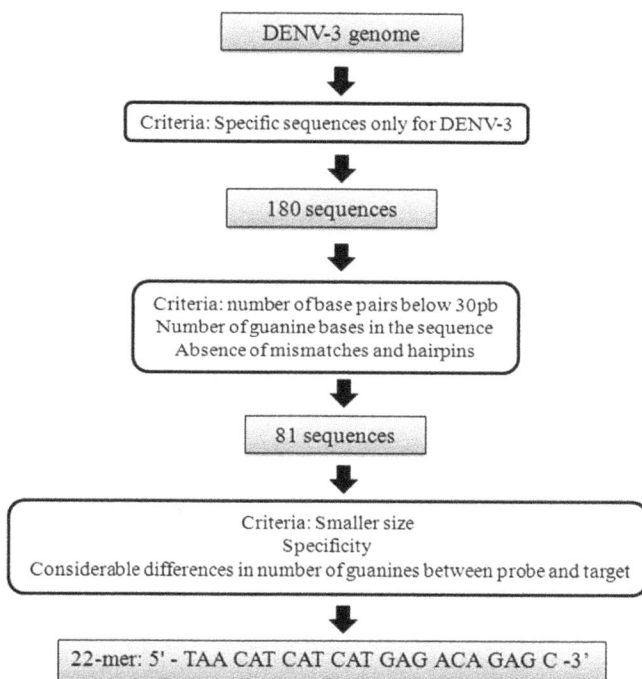

Figure 1. Flowchart of the selection criteria used to design the DENV-3 probe.

First, it was shown that 180 DNA sequences were specific only for dengue virus serotype 3. Among these, 81 were selected as DNA probes, based on the number of base pairs of the sequence, number of guanine bases and the absence of mismatches and hairpins. Finally, a 22-m oligonucleotide was selected to detect DENV-3, with the following sequence: 5′-TAA CAT CAT CAT GAG ACA GAG C-3′. This sequence was selected due to its suitable features that are desirable for electrochemical biosensors, such as shorter base pair length, high specificity and a considerable difference in the number of guanine base between the probe and target sequences, which is important to discriminate between ssDNA and dsDNA onto the PGE surface [33,39,40].

175

In addition, this probe was targeted to detect sequences from the envelope (E) gene, which is responsible for binding and fusion to host cell membranes [4,41]. This particular gene was chosen because of it is highly conserved sequence, which suffers less mutation process rather than other parts of dengue genome. Viral gene regions that interact with specific host cells are evolutionarily constrained, mainly in viruses that infect multiple organisms, like dengue virus. This is important to be considered in the development of DNA biosensors to detect dengue virus, once that it determines the selectivity and specificity of the method, avoiding cross-reactivity with non-related organisms [42–44].

3.2. Effect of DENV-3 Probes Concentration on Immobilization on the PGE

The immobilization of a biological element on the electrode is the first step to be considered in the development of a biosensor [45]. Determination of the optimal probe concentration is crucial to ensure a high performance of DNA biosensors, and reduce any interference in the electrochemical response of the system [46,47]. Thus, the effect of DENV-3 probes concentration was also investigated in this study.

Figure 2 shows current peaks of different DENV-3 probe concentrations on the PGE surface. As the electrochemical analysis in this study relies on label-free oxidation of guanine bases, the acquisition of higher current signals for DNA probes is well-suited for this system [29,48–50]. The results show that the current gradually rises with the increase of the probe concentration from 250 nM up to 500 nM, reaching the highest electrochemical signal of 777 $\pm$ 8.6 nA at 500 nM. The result obtained at 500 nM was also statistically different from that obtained at 750 nM ($p = 0.000178$). However, the decrease in the current peaks at higher concentrations of DNA probes after 500 nM could be due to the steric hindrance between the nitrogenous bases and the transducer. This prevents the electrons produced by the oxidation process to access the electrode surface [51–55]. Therefore, a concentration of 500 nM was selected as the optimal probe concentration for DNA immobilization on the PGE.

3.3. Electrochemical Analysis of Hybridization Assays

In this study, the biosensor performance was analyzed through the hybridization reaction between the DENV-3 probe and the complementary DENV-3 oligonucleotide. Hybridization was carried out with different amounts of the target sequence and this reaction was performed in an electrochemical cell containing 20 mM Tris-HCl buffer (pH 7.0). The electrochemical signals based on guanine oxidation are displayed in Figure 3. The results showed that the current peaks increase with the increasing concentration of the target sequence (10 nM to 500 nM); the highest concentration exhibited the highest current peak of the system (135 $\pm$ 2.15 nA). However, at concentrations higher than 500 nM, there is a decrease in the electrochemical signal that could be due to electrostatic hindrance of DNA molecules on the PGE surface [39,53,56].

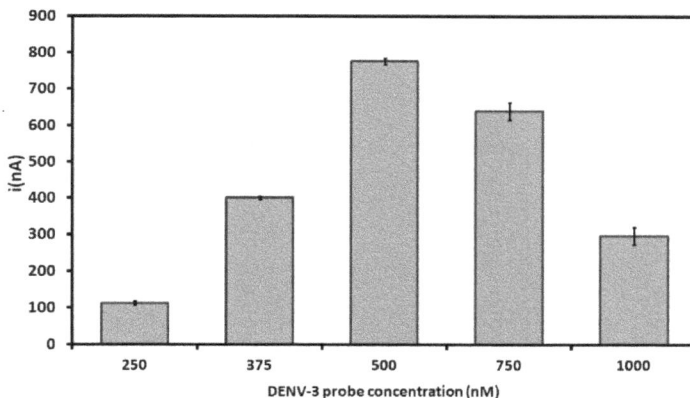

Figure 2. Electrochemical signals of different concentrations of DENV-3 probe onto pencil graphite electrodes (PGE). Differential pulse voltammetry (DPV) was used for electrochemical analysis based on guanine oxidation. Experimental conditions: Scanning potential range between +0.5 V and +1.2 V and scan rate of 0.05 V/s. The results represent the average of triplicates carried out at each DENV-3 probe concentration.

The linear regression of the current peaks obtained from different concentrations of DENV-3 target is shown in the inset of Figure 3. The calibration curve (described by the equation $y = 0.8962x + 24.979$) is linear between 10 nM and 100 nM, with a correlation coefficient of 0.9883 ($p < 0.00536$, $n = 5$). A detection limit of 3.09 nM could be estimated with the following equation: $3\,s/m$, where s is the standard deviation of most reproducible current peak result (corresponding to 75 nM concentration) and m is the slope of the linear regression [57]. The same experimental conditions were used to estimate the reproducibility of the method, which was 1.01%, indicating, thus, the significant reproducibility of the method.

Figure 4 displays electrochemical signals of probe-modified PGE before and after hybridization with 250 nM of the target. A decrease of 83% in the current signal was observed after the reaction with the DENV-3 target; this is due to the fact that oxidizable regions of guanine bases in the ssDNA (777 nA) are bound through hydrogen bonds that held the double chain together, thus decreasing the electrochemical signal of the dsDNA (135 nA) on the electrode surface [24,58–60].

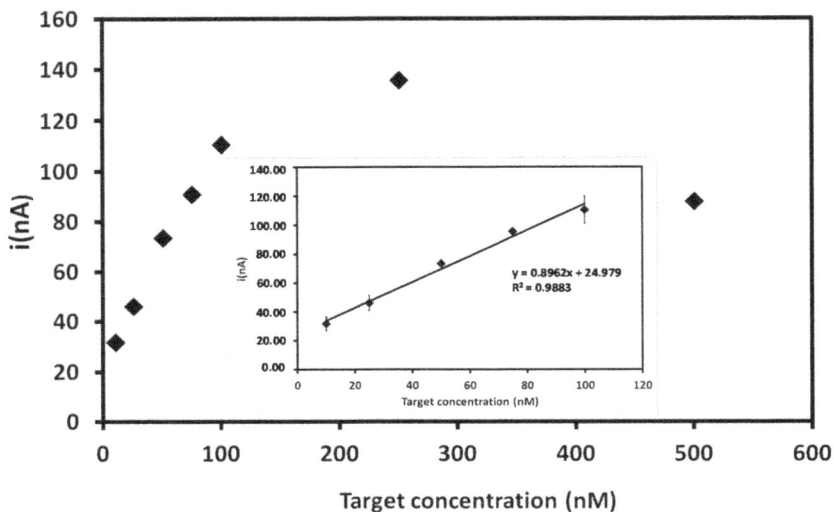

Figure 3. Current signals obtained for different DENV-3 target sequence concentrations after hybridization with probe-modified PGEs. Inset: Related calibration graph at a concentration range of 10–100 nM for the target sequence. Experimental conditions: Scanning potential range between +0.5 V and +1.2 V and scan rate of 0.05 V/s.

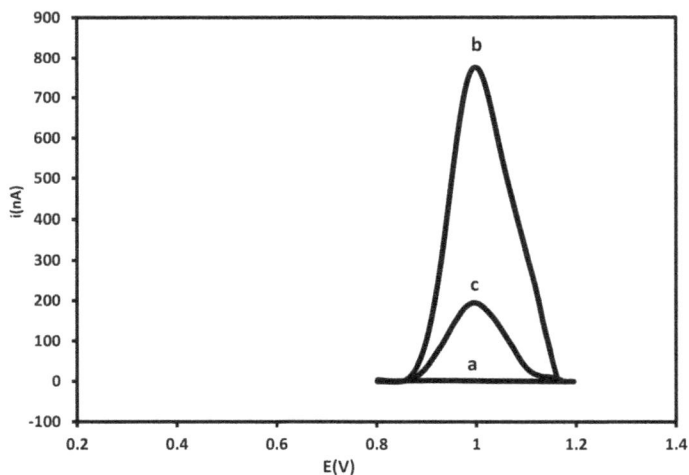

Figure 4. Differential pulse voltammograms corresponding to bare PGE (**a**), probe-modified PGE before (**b**) and after hybridization with 250 nM of target sequence (**c**) in 20mM Tris-HCl buffer solution (pH 7.0). Experimental conditions: Scanning potential range between +0.5 V and +1.2 V and scan rate of 0.05 V/s.

178

Furthermore, as is seen in Table 1, the present sensor has a lower detection limit (3.09 nM) compared to other electrochemical DNA biosensors.

Table 1. Comparison of the analytical performance of different electrochemical DNA biosensors.

Nucleic Acid Biosensor	Electrode	Electrochemical Method	Linear Range of Hybridization	Detection Limit	Reference
Single-walled carbon nanotubes-polymer modified graphite electrodes for DNA hybridization	PGE [a]	DPV [d]	50–200μg/mL	5.14 μM	[61]
Hybridization biosensor for detection of hepatitis B virus	GCE [b]	DPV	0.36–1.32 μM	19.4 nM	[62]
Brilliant cresyl blue as electroactive indicator in electrochemical DNA oligonucleotide sensors	CPE [c]	DPV	10 nM–5μM	9 nM	[63]
Label-free DNA detection based on zero current potentiometry	PGE	LSV [e]	10 nM–1μM	6.9 nM	[64]
DNA biosensor detection of DENV-3 sequences onto PGE surfaces	PGE	DPV	10–100 nM	3.09 nM	This work

[a] Pencil graphite electrode; [b] Glassy carbon electrode; [c] Carbon paste electrode; [d] Differential pulse voltammetry; [e] Linear sweep voltammetry.

3.4. Selectivity Study

In a way to evaluate the selectivity of the DENV-3 biosensor, hybridization tests were performed with a non-complementary sequence. DPV voltammograms for bare PGE, probe-modified PGE before and after hybridization with DENV-3 target and non-complementary sequence are displayed in Figure 5. It was verified that no electrochemical signal was recorded with bare PGE, which is in agreement with the absence of DNA on the electrode surface. Probe-modified PGE presented the highest current peak of the system, whereas the probe-modified PGE after hybridization with target sequence showed a decrease in the current signal, as discussed previously.

As shown in Figure 5, a significant difference in the voltammetric signal was observed after hybridization of DENV-3 probe with the non-complementary sequence (600 nA) when compared with the complementary DNA (135 nA); however, the signal was slightly lower compared to the probe-modified electrode (777 nA). This may be attributed to some unspecific hybridization of non-complementary sequences with the probe. Nevertheless, the target sequence is clearly able to form a steady dsDNA on the electrode surface. Moreover, a decrease in the current peak was also noticed when the probe-modified PGE was added to the mixed DNA solution (~230 nA) when compared with the probe-modified electrode. Therefore, these results can confirm the ability of the PGE-modified biosensor to detect selectively dengue virus serotype 3 [34,61,62].

179

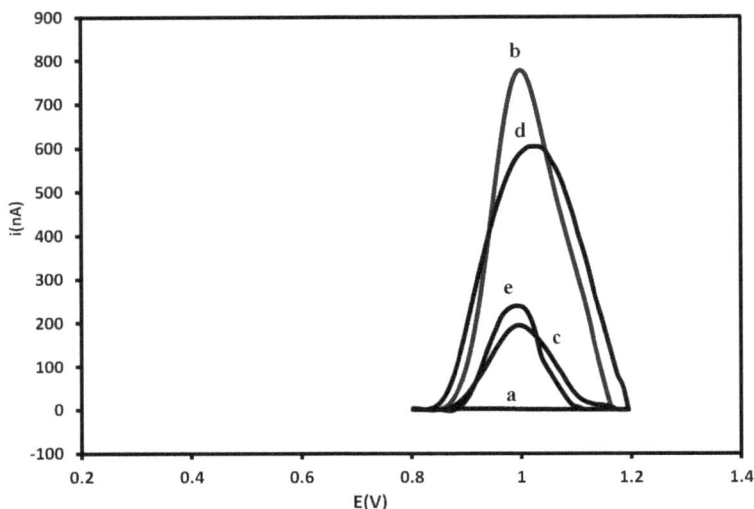

Figure 5. Differential pulse voltammograms for guanine oxidation of (**a**) bare PGE;
(**b**) probe-modified PGE; (**c**) probe-modified PGE after hybridization with DENV-3
sequence; (**d**) non-complementary sequence and (**e**) a mixed solution of DENV-3
sequence and non-complementary sequence. Experimental conditions: Scanning
potential range between +0.5 V and +1.2 V and scan rate of 0.05 V/s.

3.5. Electrochemical Measurement of Target Hybridization in Human Serum Solutions

In order to evaluate the efficiency of the probe surface for biosensing applications
and in an attempt to test the performance if the biosensor for the detection
DENV-3 in real samples, DPV was used to investigate DNA hybridization on
PGE surface using human serum. This assay was tested with 75 nM of target
sequence, non-complementary sequences and a solution mixed of both target and
non-complementary sequences.

As shown in Figure 6, the biosensor displays the same electrochemical
behavior observed previously in Tris-HCl buffer solutions. However, all the current
signals of the probe-modified PGE after hybridization with the complementary,
non-complementary sequences and mixed DNA solution diluted in human serum
presented a slight decrease (134, 410.8 and 221 nA, respectively) when compared
with those diluted in acetate buffer (135, 600 and 230 nA, respectively). This could be
due to the hybridization kinetics and the efficiency of the PGE surface, which could
be affected by non-specific adsorption of plasma proteins, and this may interfere
with the detection of the electrochemical signal [63]. However, such interference
with the detection of DNA molecules was observed previously with the optical DNA
biosensor developed by Gong *et al.* [64,65]. Thus, these results confirm the high
selectivity and sensitivity of the electrochemical DNA biosensor developed herein.

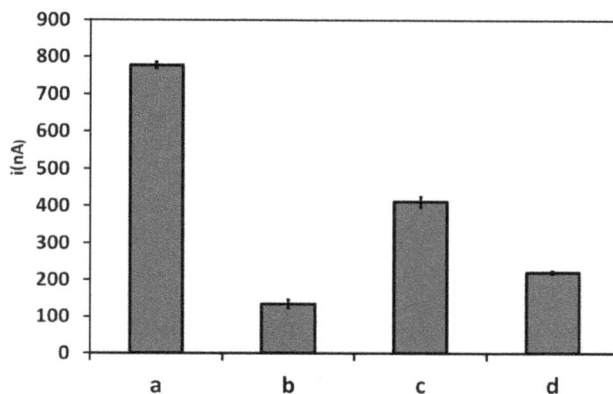

Figure 6. Current peaks related to guanine oxidation of the probe-modified-PGE after (**a**) and before hybridization with DENV-3 (**b**); in the presence of non-complementary sequences (**c**) and in a solution mixed with DENV-3 and non-complementary sequences (**d**), all diluted in human serum. Experimental conditions: Scanning potential range between +0.5 V and +1.2 V and scan rate of 0.05 V/s.

4. Conclusions

A sensitive DNA biosensor based on electrochemistry for the detection of dengue virus serotype 3 was proposed in the present study. A pencil graphite electrode, modified with a probe designed specifically for DENV-3, was able to identify selectively target sequences of the virus, with a low detection limit of 3.09 nM. Moreover, the probe-modified PGE allowed to detect specifically complementary sequences of the target DNA spiked with human serum.

The sensitivity of this assay can be further improved by testing other electrode materials, such as gold, platinum and grapheme electrodes. In addition, screen-printed electrodes could be also used for the implementation of a portable system. Therefore, the application of biosensors for the diagnosis of dengue virus at the molecular level may contribute to the future development and advancement of effective and rapid detection methods.

Acknowledgments: The authors would like to thank Centro de Pesquisas Aggeu Magalhães (CPqAM) that kindly provided biological samples for this project. In addition, they would like to thank Fundação de Amparo à Pesquisa do Estado de Pernambuco (FACEPE), Conselho Nacional de Desenvolvimento Científico e Tecnológico (CNPq) and Laboratório de Imunopatologia Keizo Asami (LIKA) for providing financial support and Graziella El Khoury for reviewing of the manuscript for English language.

Author Contributions: N.O. designed and performed the majority of the experiments. D.Z. performed bioinformatics analysis. W.B. and M.B. prepared PGEs, buffer solutions and performed part of electrochemical experiments. M.T. and K.T. were responsible for selecting and supplying biological samples for the experiments. D.M. processed human serum samples. G.N. and D.F. were responsible for statistical data analysis. N.O., E.S., M.A. and J.L.-F. drafted and revised critically the manuscript.

Conflicts of Interest: The authors declare no conflict of interest.

References

1. WHO. *Dengue—Guidelines for Diagnosis, Treatment, Prevention and Control*; WHO Press: Geneva, Switzerland, 2009; p. 147.
2. Sariol, C.A.; White, L.J. Utility, limitations, and future of non-human primates for dengue research and vaccine development. *Front. Immunol.* **2014**, *5*, 452.
3. Bhatt, S.; Gething, P.W.; Brady, O.J.; Messina, J.P.; Farlow, A.W.; Moyes, C.L.; Drake, J.M.; Brownstein, J.S.; Hoen, A.G.; Sankoh, O.; Myers, M.F.; *et al.* The global distribution and burden of dengue. *Nature* **2013**, *496*, 504–507.
4. Vasilakis, N.; Cardosa, J.; Hanley, K.A.; Holmes, E.C.; Weaver, S.C. Fever from the forest: Prospects for the continued emergence of sylvatic dengue virus and its impact on public health. *Nat. Rev. Microbiol.* **2011**, *9*, 532–541.
5. Chien, L.C.; Yu, H.L. Impact of meteorological factors on the spatiotemporal patterns of dengue fever incidence. *Environ. Int.* **2014**, *73*, 46–56.
6. Cedillo-Barrón, L.; García-Cordero, J.; Bustos-Arriaga, J.; León-Juárez, M.; Gutiérrez-Castañeda, B. Antibody response to dengue virus. *Microbes Infect.* **2014**, *16*, 711–720.
7. Rovida, F.; Percivalle, E.; Campanini, G.; Piralla, A.; Novati, S.; Muscatello, A.; Baldanti, F. Viremic Dengue virus infections in travellers: Potential for local outbreak in Northern Italy Dengue-1 Dengue-4. *J. Clin. Virol.* **2011**, *50*, 76–79.
8. Ashley, E.A. Trends in Anaesthesia and Critical Care Dengue fever. *Trends Anaesth. Crit. Care* **2011**, *1*, 39–41.
9. Qing, X.; Sun, N.; Yeh, J.; Yue, C.; Cai, J. Dengue fever and bone marrow myelofibrosis. *Exp. Mol. Pathol.* **2014**, *97*, 208–210.
10. Halstead, S.B. Dengue. *Lancet* **2007**, *370*, 1644–1652.
11. Arora, P.; Sindhu, A.; Dilbaghi, N.; Chaudhury, A. Biosensors as innovative tools for the detection of food borne pathogens. *Biosens. Bioelectron.* **2011**, *28*, 1–12.
12. Chakravarti, A.; Arora, R.; Luxemburger, C. Fifty years of dengue in India. *Trans. R. Soc. Trop. Med. Hyg.* **2012**, *106*, 273–282.
13. Verma, R.; Sahu, R.; Holla, V. Neurological manifestations of dengue infection: A review. *J. Neurol. Sci.* **2014**, *346*, 26–34.
14. Pechansky, F.; Duarte, P.D.C.A.V.; de Boni, R.; Leukefeld, C.G.; von Diemen, L.; Bumaguin, D.B.; Kreische, F.; Hilgert, J.B.; Bozzetti, M.C.; Fuchs, D.F.P. Predictors of positive Blood Alcohol Concentration (BAC) in a sample of Brazilian drivers. *Rev. Bras. Psiquiatr.* **2012**, *34*, 277–285.

15. Ferraz, F.O.; Bomfim, M.R.Q.; Totola, A.H.; Ávila, T.V.; Cisalpino, D.; Pessanha, J.E.M.; da Glória de Souza, D.; Teixeira Júnior, A.L.; Nogueira, M.L.; Bruna-Romero, O.; *et al.* Evaluation of laboratory tests for dengue diagnosis in clinical specimens from consecutive patients with suspected dengue in Belo Horizonte, Brazil. *J. Clin. Virol.* **2013**, *58*, 41–46.

16. Peeling, R.W.; Artsob, H.; Pelegrino, J.L.; Buchy, P.; Cardosa, M.J.; Devi, S.; Enria, D.A.; Farrar, J.; Gubler, D.J.; Guzman, M.G.; *et al.* Evaluation of diagnostic tests: Dengue. *Nat. Rev. Microbiol.* **2010**, *8*, S30–S37.

17. Wattal, C.; Goel, N. Infectious Disease Emergencies in Returning Travelers—Special Reference to Malaria, Dengue Fever and Chikungunya. *Med. Clin. North Am.* **2012**, *96*, 1225–1255.

18. Korhonen, E.M.; Huhtamo, E.; Virtala, A.M.K.; Kantele, A.; Vapalahti, O. Approach to non-invasive sampling in dengue diagnostics: Exploring virus and NS1 antigen detection in saliva and urine of travelers with dengue. *J. Clin. Virol.* **2014**, *61*, 353–358.

19. Shenoy, B.; Menon, A.; Biradar, S. Science Direct Diagnostic utility of dengue NS1 antigen. *Pediatr. Infect. Dis.* **2014**, *6*, 110–113.

20. Hapugoda, M.D.; de Silva, N.R.; Khan, B.; Damsiri Dayanath, M.Y.; Gunesena, S.; Prithimala, L.D.; Abeyewickreme, W. A comparative retrospective study of RT-PCR-based liquid hybridization assay for early, definitive diagnosis of dengue. *Trans. R. Soc. Trop. Med. Hyg.* **2010**, *104*, 279–282.

21. Back, A.T.; Lundkvist, A. Dengue viruses—An overview. *Infect. Ecol. Epidemiol.* **2013**, *1*, 1–21.

22. Siddiquee, S.; Rovina, K.; Yusof, N.A.; Rodrigues, K.F. Nanoparticle-enhanced electrochemical biosensor with DNA immobilization and hybridization of Trichoderma harzianum gene. *Sens. Bio-Sens. Res.* **2014**, *2*, 16–22.

23. Teles, F.S.R.R. Biosensors and rapid diagnostic tests on the frontier between analytical and clinical chemistry for biomolecular diagnosis of dengue disease: A review. *Anal. Chim. Acta* **2011**, *687*, 28–42.

24. Lucarelli, F.; Tombelli, S.; Minunni, M.; Marrazza, G.; Mascini, M. Electrochemical and piezoelectric DNA biosensors for hybridisation detection. *Anal. Chim. Acta* **2008**, *609*, 139–159.

25. Souada, M.; Piro, B.; Reisberg, S.; Anquetin, G.; Noël, V.; Pham, M.C. Label-free electrochemical detection of prostate-specific antigen based on nucleic acid aptamer. *Biosens. Bioelectron.* **2014**, *68C*, 49–54.

26. Tosar, J.P.; Keel, K.; Laíz, J. Two independent label-free detection methods in one electrochemical DNA sensor. *Biosens. Bioelectron.* **2009**, *24*, 3036–3042.

27. Sadik, O.A.; Aluoch, A.O.; Zhou, A. Status of biomolecular recognition using electrochemical techniques. *Biosens. Bioelectron.* **2009**, *24*, 2749–2765.

28. Conde, J.; Edelman, E.R.; Artzi, N. Target-responsive DNA/RNA nanomaterials for microRNA sensing and inhibition: The jack-of-all-trades in cancer nanotheranostics? *Adv. Drug Deliv. Rev.* **2015**, *81C*, 169–183.

29. Souza, E.; Nascimento, G.; Santana, N.; Ferreira, D.; Lima, M.; Natividade, E.; Martins, D.; Lima-Filho, J. Label-free electrochemical detection of the specific oligonucleotide sequence of dengue virus type 1 on pencil graphite electrodes. *Sensors* **2011**, *11*, 5616–5629.

30. Lourenço-de-Oliveira, R.; Honório, N.A.; Castro, M.G.; Schatzmayr, H.G.; Miagostovich, M.P.; Alves, J.C.R.; Silva, W.C.; Leite, P.J.; Nogueira, R. Dengue Virus Type 3 Isolation from Aedes aegypti in the Municipality of Nova Iguaçu, State of Rio de Janeiro. *Mem. Inst. Oswaldo Cruz* **2002**, *97*, 799–800.

31. Peyrefitte, C.N.; Couissinier-Paris, P.; Mercier-Perennec, V.; Bessaud, M.; Martial, J.; Kenane, N.; Durand, J.P.A.; Tolou, H.J. Genetic Characterization of Newly Reintroduced Dengue Virus Type 3 in Martinique (French West Indies). *J. Clin. Microbiol.* **2003**, *41*, 5195–5198.

32. Ensafi, A.A.; Heydari-bafrooei, E.; Amini, M. DNA-functionalized biosensor for riboflavin based electrochemical interaction on pretreated pencil graphite electrode. *Biosens. Bioelectron.* **2012**, *31*, 376–381.

33. Hejazi, M.S.; Alipour, E.; Pournaghi-Azar, M.H. Immobilization and voltammetric detection of human interleukine-2 gene on the pencil graphite electrode. *Talanta* **2007**, *71*, 1734–1740.

34. Pournaghi-Azar, M.H.; Alipour, E.; Zununi, S. Direct and rapid electrochemical biosensing of the human interleukin-2 DNA in unpurified polymerase chain reaction (PCR)-amplified real samples. *Biosens. Bioelectron.* **2008**, *24*, 524–530.

35. Özcan, A.; Yücel, S. A novel approach for the determination of paracetamol based on the reduction of N-acetyl-p-benzoquinoneimine formed on the electrochemically treated pencil graphite electrode. *Anal. Chim. Acta* **2011**, *685*, 9–14.

36. Ermini, M.L.; Scarano, S.; Bini, R.; Banchelli, M.; Berti, D.; Mascini, M.; Minunni, M. A rational approach in probe design for nucleic acid-based biosensing. *Biosens. Bioelectron.* **2011**, *26*, 4785–4790.

37. Huang, J.; Yang, X.; He, X.; Wang, K.; Liu, J.; Shi, H.; Wang, Q.; Guo, Q.; He, D. Design and bioanalytical applications of DNA hairpin-based fluorescent probes. *TrAC Trends Anal. Chem.* **2014**, *53*, 11–20.

38. O'Brien, B.; Zeng, H.; Polyzos, A.A.; Lemke, K.H.; Weier, J.F.; Wang, M.; Zitzelsberger, H.F.; Weier, H.U.G. Bioinformatics tools allow targeted selection of chromosome enumeration probes and aneuploidy detection. *J. Histochem. Cytochem.* **2013**, *61*, 134–147.

39. Campos-Ferreira, D.S.; Souza, E.; Nascimento, G.; Zanforlin, D.; Arruda, M.; Beltrão, M.; Melo, A.; Bruneska, D.; Lima-Filho, J.L. Electrochemical DNA biosensor for the detection of human papillomavirus E6 gene inserted in recombinant plasmid. *Arab. J. Chem.* **2014**.

40. Corrigan, D.K.; Schulze, H.; McDermott, R.A.; Schmüser, I.; Henihan, G.; Henry, J.B.; Bachmann, T.T.; Mount, A.R. Improving electrochemical biosensor performance by understanding the influence of target DNA length on assay sensitivity. *J. Electroanal. Chem.* **2014**, *732*, 25–29.

41. Soares, R.O.S.; Caliri, A. Stereochemical features of the envelope protein Domain III of dengue virus reveals putative antigenic site in the five-fold symmetry axis. *Biochim. Biophys. Acta* **2013**, *1834*, 221–230.

42. Weaver, S.C.; Brault, A.C.; Kang, W.; John, J. Genetic and Fitness Changes Accompanying Adaptation of an Arbovirus to Vertebrate and Invertebrate Cells Genetic and Fitness Changes Accompanying Adaptation of an Arbovirus to Vertebrate and Invertebrate Cells. *J. Virol.* **1999**, *73*, 4316–4326.

43. Bennett, S.N.; Holmes, E.C.; Chirivella, M.; Rodriguez, D.M.; Beltran, M.; Vorndam, V.; Gubler, D.J.; McMillan, W.O. Molecular evolution of dengue 2 virus in Puerto Rico: Positive selection in the viral envelope accompanies clade reintroduction. *J. Gen. Virol.* **2006**, *87*, 885–893.

44. Weaver, S.C.; Vasilakis, N. Molecular evolution of dengue viruses: Contributions of phylogenetics to understanding the history and epidemiology of the preeminent arboviral disease. *Infect. Genet. Evol.* **2009**, *9*, 523–540.

45. Wang, Q.; Ding, Y.; Gao, F.; Jiang, S.; Zhang, B.; Ni, J.; Gao, F. A sensitive DNA biosensor based on a facile sulfamide coupling reaction for capture probe immobilization. *Anal. Chim. Acta* **2013**, *788*, 158–164.

46. Zhang, L.; Wang, Y.; Chen, M.; Luo, Y.; Deng, K.; Chen, D.; Fu, W. A new system for the amplification of biological signals: RecA and complimentary single strand DNA probes on a leaky surface acoustic wave biosensor. *Biosens. Bioelectron.* **2014**, *60*, 259–264.

47. Mohamadi, M.; Mostafavi, A.; Torkzadeh-Mahani, M. Electrochemical determination of biophenol oleuropein using a simple label-free DNA biosensor. *Bioelectrochemistry* **2015**, *101C*, 52–57.

48. Erdem, A.; Muti, M.; Karadeniz, H.; Congur, G.; Canavar, E. Colloids and Surfaces B: Biointerfaces Electrochemical monitoring of indicator-free DNA hybridization by carbon nanotubes—Chitosan modified disposable graphite sensors. *Colloids Surf. B Biointerfaces* **2012**, *95*, 222–228.

49. Paleček, E.; Fojta, M.; Tomschik, M.; Wang, J. Electrochemical biosensors for DNA hybridization and DNA damage. 1998, 13, 621–628.

50. Wang, J. Electrochemical biosensors: Towards point-of-care cancer diagnostics. *Biosens. Bioelectron.* **2006**, *21*, 1887–92.

51. Campos-Ferreira, D.S.; Nascimento, G.A.; Souza, E.V.M.; Souto-maior, M.A.; Arruda, M.S.; Zanforlin, D.M.L.; Ekert, M.H.F.; Bruneska, D.; Lima-filho, J.L. Electrochemical DNA biosensor for human papillomavirus 16 detection in real samples. *Anal. Chim. Acta* **2013**, *804*, 258–263.

52. Gao, Z.; Yang, W.; Wang, J.; Yan, H.; Yao, Y.; Ma, J.; Wang, B.; Zhang, M.; Liu, L. Electrochemical synthesis of layer-by-layer reduced graphene oxide sheets/polyaniline nanofibers composite and its electrochemical performance. *Electrochim. Acta* **2013**, *91*, 185–194.

53. Lucarelli, F.; Marrazza, G.; Palchetti, I.; Cesaretti, S.; Mascini, M. Coupling of an indicator-free electrochemical DNA biosensor with polymerase chain reaction for the detection of DNA sequences related to the apolipoprotein E. *Anal. Chim. Acta* **2002**, *469*, 93–99.

54. Teles, F.R.R.; Fonseca, L.P. Trends in DNA biosensors. *Talanta* **2008**, *77*, 606–623.

55. Tichoniuk, M.; Ligaj, M.; Filipiak, M. Application of DNA Hybridization Biosensor as a Screening Method for the Detection of Genetically Modified Food Components. *Sensors* **2008**, *8*, 2118–2135.

56. Liu, X.; Fan, Q.; Huang, W. DNA biosensors based on water-soluble conjugated polymers. *Biosens. Bioelectron.* **2011**, *26*, 2154–2164.

57. Gumustas, M.; Ozkan, S.A. The Role of and the Place of Method Validation in Drug Analysis Using Electroanalytical Techniques. *Open Anal. Chem. J.* **2011**, *5*, 1–21.

58. Nascimento, G.A.; Souza, E.V.M.; Campos-ferreira, D.S.; Arruda, M.S.; Castelletti, C.H.M.; Wanderley, M.S.O.; Ekert, M.H.F.; Bruneska, D. Electrochemical DNA biosensor for bovine papillomavirus detection using polymeric film on screen-printed electrode. *Biosens. Bioelectron.* **2012**, *38*, 61–66.

59. Aydoğdu, G.; Günendi, G.; Zeybek, D.K.; Zeybek, B.; Pekyardımcı, Ş. A novel electrochemical DNA biosensor based on poly-(5-amino-2-mercapto-1,3,4-thiadiazole) modified glassy carbon electrode for the determination of nitrofurantoin. *Sens. Actuators B Chem.* **2014**, *197*, 211–219.

60. Wang, J.; Kawde, A. Pencil-based renewable biosensor for label-free electrochemical detection of DNA hybridization. *Anal. Chim. Acta* **2001**, *431*, 219–224.

61. Sehatnia, B.; Golabi, F.; Sabzi, R.E.; Hejazi, M.S. Modeling of DNA Hybridization Detection Using Methylene Blue as an Electroactive Label. *J. Iran. Chem. Soc.* **2011**, *8*, 115–122.

62. Souza, E.; Nascimento, G.; Santana, N.; Campos-ferreira, D.; Bibiano, J.; Arruda, M.S. Electrochemical DNA Biosensor for Sequences Related to the Human Papillomavirus Type 16 using Methylene Blue. *Biosens. J.* **2014**, *3*, 3–7.

63. Ren, Y.; Deng, H.; Shen, W.; Gao, Z. A Highly Sensitive and Selective Electrochemical Biosensor for Direct Detection of MicroRNAs in Serum. *Anal. Chem.* **2013**, *9*, 4784–4789.

64. Auer, S.; Nirschl, M.; Schreiter, M.; Vikholm-Lundin, I. Detection of DNA hybridisation in a diluted serum matrix by surface plasmon resonance and film bulk acoustic resonators. *Anal. Bioanal. Chem.* **2011**, *400*, 1387–1396.

65. Gong, P.; Lee, C.; Gamble, L.J.; Castner, D.G.; Grainger, D.W.; Chem, L.J. A. Hybridization Behavior of Mixed DNA/Alkylthiol Monolayers on Gold: Characterization by Surface Plasmon Resonance and 32 P Radiometric Assay rescence intensity measurements reported in a related. **2006**, *78*, 3326–3334.

An Apta-Biosensor for Colon Cancer Diagnostics

Mojgan Ahmadzadeh Raji, Ghasem Amoabediny, Parviz Tajik, Morteza Hosseini and Ebrahim Ghafar-Zadeh

Abstract: This paper reports the design and implementation of an aptasensor using a modified KCHA10a aptamer. This aptasensor consists of a functionalized electrodes using various materials including 11-mercaptoandecanoic acid (11-MUA) and modified KCHA10a aptamer. The HCT 116, HT 29 and HEp-2 cell lines are used in this study to demonstrate the functionality of aptasensor for colon cancer detection purposes. Flow cytometry, fluorescence microscopy and electrochemical cyclic voltammetry are used to verify the binding between the target cells and aptamer. The limit of detection (LOD) of this aptasensor is equal to seven cancer cells. Based on the experimental results, the proposed sensor can be employed for point-of-care cancer disease diagnostics.

Reprinted from *Sensors*. Cite as: Raji, M.A.; Amoabediny, G.; Tajik, P.; Hosseini, M.; Ghafar-Zadeh, E. An Apta-Biosensor for Colon Cancer Diagnostics. *Sensors* **2015**, *15*, 22291–22303.

1. Introduction

Colorectal cancer is the second and third most common cause of cancer deaths in Canada and Iran, respectively [1,2]. Rapid diagnosis of this disease increases the chance of survival and decreases the medical management cost. Aptasensors have attracted attention for potential point-of-care diagnostic applications of a variety of deadly diseases such as prostate and colorectal cancers [3–5]. In these sensors, aptamers immobilized on the surface of electrodes play a key role as a recognition element for the detection of biomarkers associated with the various diseases. The interactions between aptamers and target cells/molecules are measured using various techniques, including optical and electrochemical ones [6–8]. The focus of this paper is on the design and implementation of an electrochemical aptasensor for colon cancer detection, as shown in Figure 1.

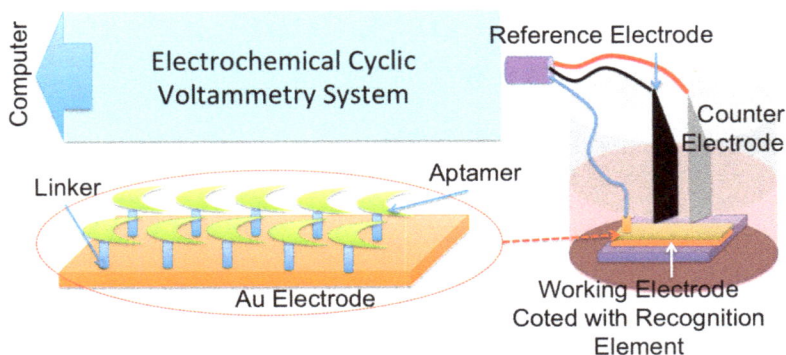

Figure 1. Illustration of an apta-sensor using the electrochemical reading technique.

Since the late 1970s, carcinoembryonic antigen (CEA) has used as a major biomarker for the detection of colon cancer and other tumors of epithelial origin. CEA is a highly glycosylated biomarker which is expressed at the surface of the HCT 116 human cell line. This antigen provides selective high affinity binding to aptamers synthesized for colon cancer detection. Anti-CEA antibody can also be considered as an alternative for the detection of colon cancer [9–11]; however, in contrast to antibodies, aptamers offer the key advantages of greater specificity and affinity binding with target molecules. Indeed the small size of aptamers (<100 nucleotides [12]) in comparison to antibodies (~10 nm [13]), is the key factor in creating such high affinity and sensitivity. Further, aptamers are more stable under ambient conditions (e.g., temperature = 25°) than antibodies, and thus have a much longer shelf life. Additionally aptamers are produced through economical chemical process from large compound libraries containing different sequences using the systematic evolution of ligands by exponential enrichment (SELEX) procedure, while antibodies are routinely extracted from animals (e.g., rabbit, sheep *etc.*) [14–18]. Aptamers are self-refolding and reusable, while antibody-based biosensors are disposable. Based on the above mentioned reasons, high sensitivity and low cost aptasensors are best candidates for colon cancer screening [19].

The immobilization of aptamers on the surface of gold electrodes or gold nanoparticles [20,21] using various methods is an important step to develop a biosensor. A low complexity method for this purpose relies on the simple physical adsorption of DNA aptamers on the gold electrodes [22]. This method does not offer stable binding due to the relatively weak and unreliable van der Waals forces between the surfaces of electrodes and aptamers. On other hand, covalent chemical bonding techniques can be employed to develop stable and strong linkers. As described in the next section, self-assembled monolayers of substances such as

11-mercaptoundecanoic acid (11-MUA) are stacked with strong chemical binding to form a linker between the aptamers and working electrode [20–23].

In the remainder of this paper, we present the design and implementation of our functionalized sensor in Section 2. Thereafter, in Section 3, the materials and experimental protocols are described. In Section 4, we also demonstrate and discuss the experimental results followed by the conclusions in Section 5.

2. Aptasensor Design

In this section, we describe the experimental techniques for the development of an aptasensor dedicated to colon cancer detection. This section presents the synthesis of the aptamer and creation of recognition elements immobilized on the electrode surface.

2.1. Aptamer Synthesis

The aptamer synthesis is the first step toward the development of our biosensor. In this process, SELEX is employed to search for the aptamer DNA sequence. The following structure called KCHA10a (Figure 2a) is extracted through this process for colon cancer cell detection [19]. This aptamer sequence was functionalized by adding a linker and a label, namely amino-C6 at the 3′ end and fluorescein isothiocyanate (FITC) at the 5′ end [24]. This aptamer, called FITC-DNA Aptamer RAJI2-HP (Figure 2b), was synthesized by Tag Copenhagen Inc. (Frederiksberg, Denmark).

5'-
ACGCAGCAGGGGAGGCGAGAGCGC
ACAATAACGATGGTTGGGACCCAACT
GTTTGGACA
-3'

(a)

5'-FITC-
ACGCAGCAGGGGAGGCGAGA
GCGCACAATAACGATGGTTGG
GACCCAACTGTTTGGACA
-Amino-C6-3'

(b)

Figure 2. Synthesised DNA aptamer (**a**) prior to and (**b**) after labeling.

The FITC label is used to detect the binding between the cells and aptamer using flow cytometry and fluorescence microscopy. Different secondary structures of this aptamer can be formed, as shown in Figure 3. These models are predicated using secondary structure prediction software [25]. As seen in this figure, five different structures are predicted for the secondary structure of the KCHA10a truncated aptamer. The lower free energy (E) in these predicted structures (-5.3 kcal·mol^{-1} < E < -3.3 kcal·mol^{-1}) results in more stability in this nanosystem and consequently higher affinity to the colon cancer cells, confirmed by the 28.2 ± 2.7 nM value of Kd [12].

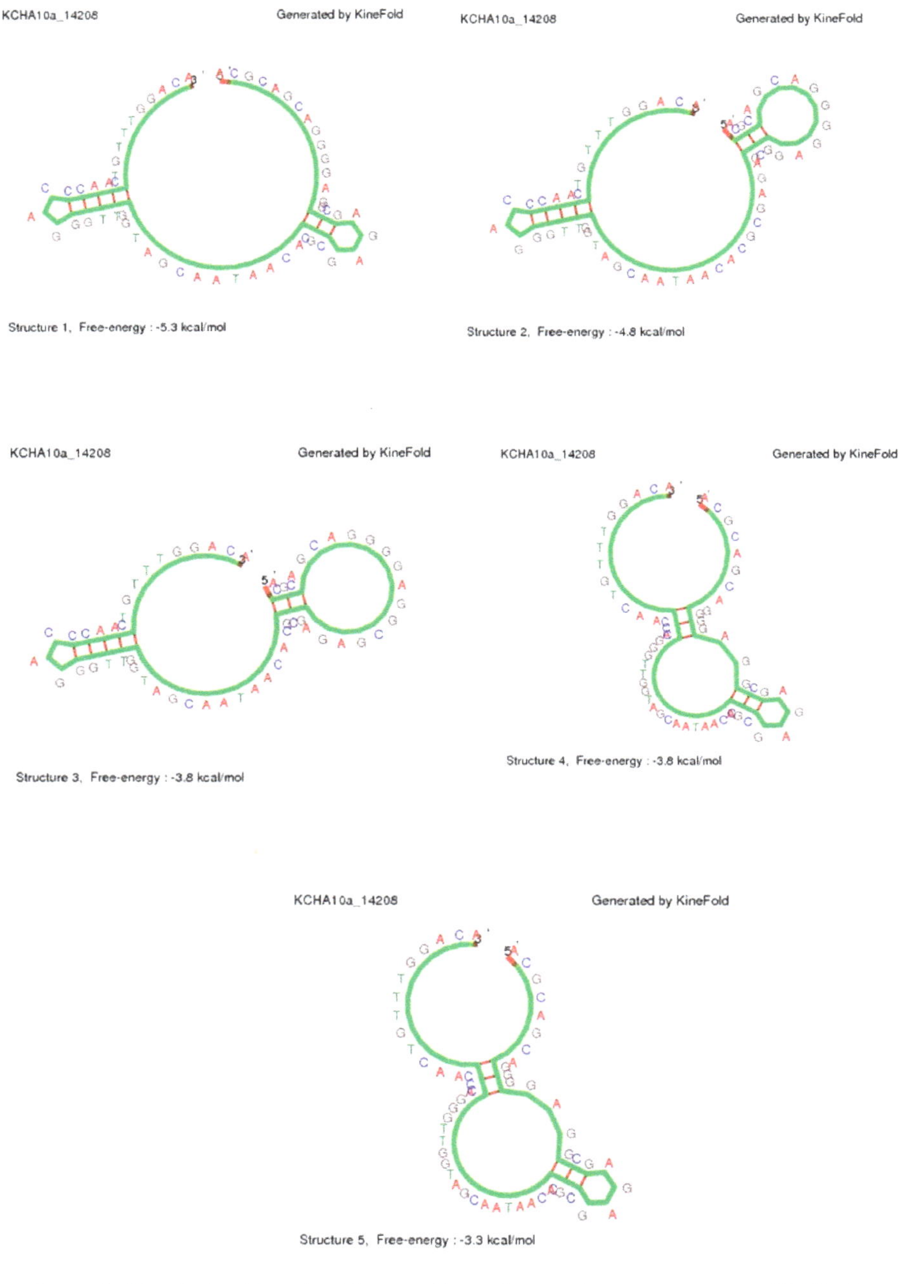

Figure 3. Simulation Results. Five predicated secondary structures of KCHA10a truncated aptamer.

2.2. Recognition Element

A chemical procedure is performed to create a recognition element consisting of linkers connected to aptamers. In this paper, the following four-step procedure was employed to create the linker:

- Creation of thiol group by coating 11-MUA on electrode.
- Activation of COOH group in 11-MUA with ethyl(dimethylaminopropyl)carbodiimide (EDC)/N-hydroxysuccinimide (NHS).
- Binding with the NH_2 at the 3′ end of the aptamer.
- Coating with bovine serum albumin (BSA) to prevent non-specific binding.

Figure 4 shows all four steps of the creation of a single linker between the surface of an Au electrode (standard) and an aptamer molecule. However, the gold surface is covered with a large number of linkers terminated with aptamers for trapping and detecting the colon cancer cells.

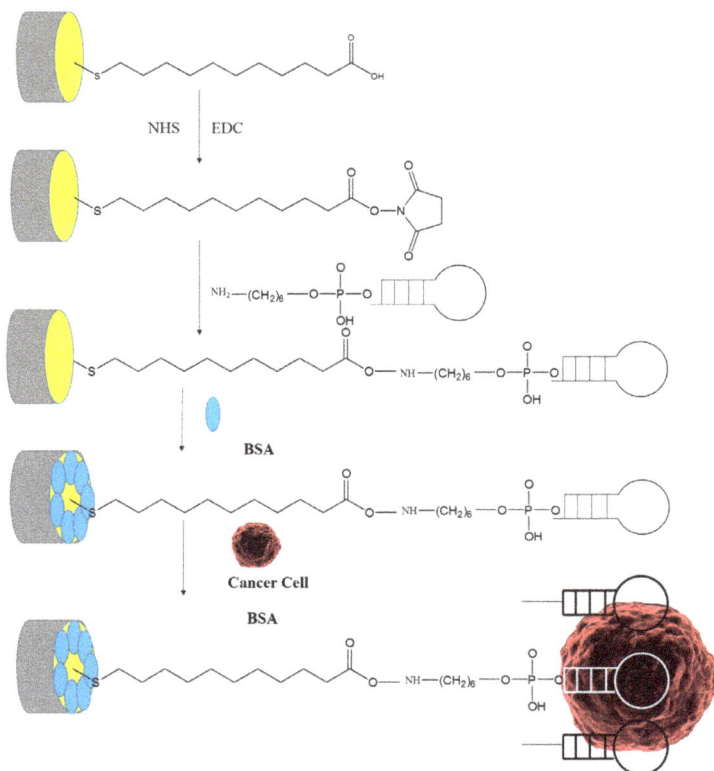

Figure 4. Schematic of the five step chemical process for the functionalization of an Au electrode dedicated for colon cancer detection.

3. Materials and Methods

3.1. Apta-Sensor Development

Gold electrode (Type: Au El., Code: 303013, Model: IRI.200-E, Azar Electrode Inc. (Orumieh, Iran) was cleaned prior to create the recognition element through four steps. In the first step, in order to coat thiol groups on the electrode surface, the electrode was soaked in ethanol/H_2O solution 3:1 (v/v) containing 20 mM 11-MUA (95%, Sigma Aldrich, London, UK) for 18 h. Then the electrode was washed with ethanol and H_2O, and dried with nitrogen gas. In the second step of this process, N-hydroxysuccinimide (NHS) and N-ethyl-N-(3-diethylaminopropyl)carbodiimide (EDC), also provided by Sigma-Aldrich, were used. EDC was used as a cross-linker to form an amide bond (Figure 4) and NHS was used to activate the carboxyl group associated with EDC. For this, the gold electrode was immersed in phosphate buffer saline (PBS) from Gibco® (London, UK) for 1 h. This buffer, with an adjusted pH at 7.4, contained 2 mM EDC and 5 mM NHS. Thereafter the electrode was soaked in Tris-HCl buffer for 2 h. This buffer with an adjusted pH at 7.6 and adjusted ionic strength I at 0.14 contained 0.4 μM aptamer. In the last step, in order to avoid non-specific binding, electrodes were dipped in distilled water containing 1% BSA for half an hour. They were also soaked in Tris-HCl buffer [20,26] for 10 min to remove the unbonded BSA.

3.2. Cell Culture

In this project, three different cells were employed to verify the functionality of the apta-sensor. The device was tested using two cells associated with human colon cancer namely epithelial cancer cell lines (Pasteur Institute of Iran, Tehran, Iran); HCT116 (NCBI code: C570) and epithelial-like cancer cell line HT 29 (NCBI code: C466). Epithelial cell HEp-2 was also used as a control cell line. Also, McCoy's 5A modified medium (Pasteur Institute of Iran, Catalogue number: 30-2007 from Gibco) was used to culture the colon cancer cells. This medium contained L-glutamine, penicillin, streptomycin, amphotericin B at concentrations of 300 mg/L, 100 μg/mL, 100 IU/mL, and 2.5 μg/mL, respectively. Also it contained fetal bovine serum, epidermal growth and adhesion factors (Nano Zist Arrayeh Inc., Tehran, Iran) along with antitrypsin activity factor to promote cell proliferation and cell attachment in the adherent flasks provided from JET BIOFII. For the control cells, RPMI (Bioeideh Inc., Tehran, Iran) was used as cell culture medium. To culture all types of cells, the incubation was performed in a SANYO device (model MCO-17AI, LabX, Tokyo, Japan) at 37 °C and carbon dioxide concentration was set to 5%. When control HEp-2 cells are confluent (25 cm^2 flasks, JET BIOFIL, Guangzhou, China) the passage of cells was performed by discarding the culture medium and trypsinizing the cells using Ethylenediaminetetraacetic acid (EDTA) solution containing 0.25% w/v trypsin

192

plus 0.53 mM EDTA [27]. However, for other cell lines, a non-enzymatic solution or cell scraper should be used instead of trypsin when the aptamer is attached to cells, because CEA is expressed on the cell surface and trypsin damages the cells. Detached cells were washed with washing buffer, centrifuged at 1200 RPM for 5 min, re-suspended in binding buffer and kept in room temperature for 20 min. An inverted microscope (IX70, Olympus, Tokyo, Japan) was used for the assessment and control of cell culture.

Two binding and washing buffers were also used to enable the conjugation of aptamers to cells with high affinity and sensitivity. Washing buffer containing 5 mM $MgCl_2$ and 4.5 g/L glucose were used for rinsing the cells after exposing the cells on the aptasensor. The binding buffer contained 1 mg/mL BSA in washing buffer [28].

3.3. Flow Cytometry & Fluorescent Microscopy

A flow cytometer (Partec, Nuremberg, Germany) and fluorescence microscope (BX50, Olympus) were employed for the assessment of binding between the aptamer and cells. For this purpose, FITC was applied as a fluorescein molecule functionalized with an isothiocyanate reactive group (–N=C=S), with excitation and emission spectrum peak wavelengths ranging from 495 to 519 nm. This spectrum range was detectable in the FL1 channel of the flow cytometer after gating and determining the desired range of negative control cells. Furthermore, as the positive and negative control of these experiments, aptamers with 400 nanomolar and zero concentrations were used, respectively.

3.4. Electrochemical Experiments

All electrochemical measurements were performed at 25 °C temperature in PBS buffer containing $K_3[Fe(CN)_6]$ with 1 mM concentration [20]. PBS adjusted at a pH equal to 7.4 was used as an electrolyte in this measurement procedure. Cyclic voltammetry (CV) was used as an electrochemical technique to detect the materials coated on the gold electrode layer by layer. These materials include 11-MUA, EDC/NHS and aptamer.

4. Results

4.1. Fluorescence Microscopy

The binding of colon cancer cells to the aptamer were studied by fluorescence microscopy using the HCT 116 and HEp-2 cell lines. Figure 5a–d shows the microscopic images of these cells prior (Figure 5a,c) and after UV light exposure (Figure 5b,d). Based on these results, the binding between the aptamer and HCT 116 cells was confirmed (Figure 5b), while no binding between the HEp-2 cells and aptamer were indicated (Figure 5d). These experiments were repeated for three times.

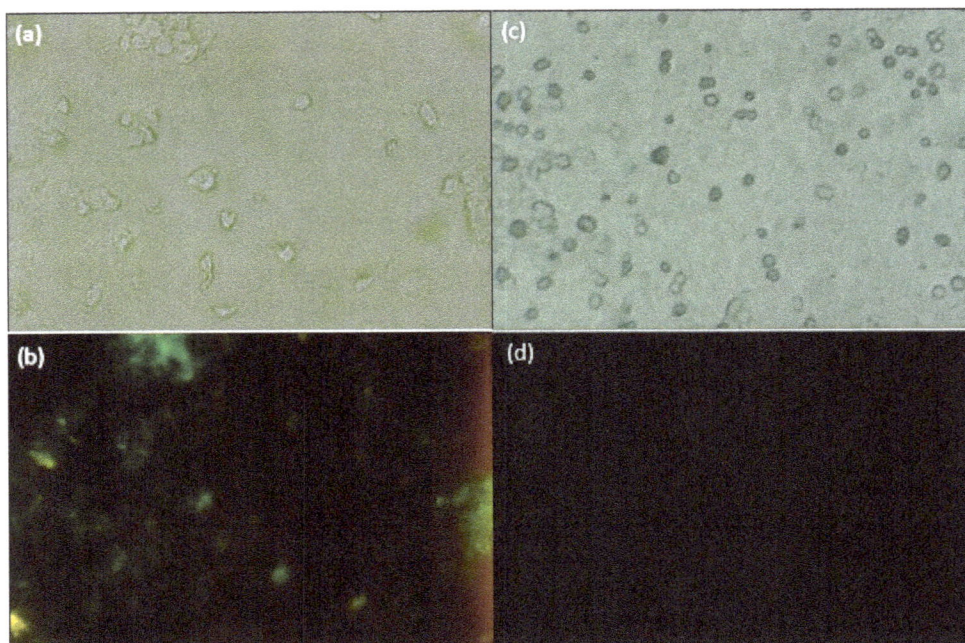

Figure 5. Fluorescence microscopy results: HCT 116 cell lines (**a**) prior and (**b**) after UV light; HEp-2 cell line (**c**) before and (**d**) after UV exposure (Olympus Fluorescence Microscope Model BX50, 40× magnification, Tokyo, Japan).

4.2. Flow Cytometry

Flow cytometry is the best technique to verify the binding between the target cells and aptamer. In this experiment, the binding between HCT 116 and HT 29 cells and aptamers was studied. As already mentioned, HEp-2 cells are used as negative control. Figure 6a–f shows the side scatter cytometry (SSC) or granularity of different types of cells in the presence and absence of aptamers as a function of the cell numbers measured by the first channel of flow cytometer (FL1). For instance, Figure 6a,b indicates that there was no difference between the surface's complexity of HEp-2 cells, before and after binding with aptamers. The significant difference between the two graphs (Figure 6c,d) reveals the interaction between the HCT 116 and aptamers. Less interaction between the HT 29 cells and aptamer is expected, as seen in Figure 6e,f. Flow cytometry histograms confirmed 60.19% and 29.62% aptamer connection to HCT 116 and HT 29 respectively. The percentage of attached cells analysed with the Flomax software is also shown in Table 1. Based on these results, the R1 region indicates the interaction with cells before and after introducing aptamers.

194

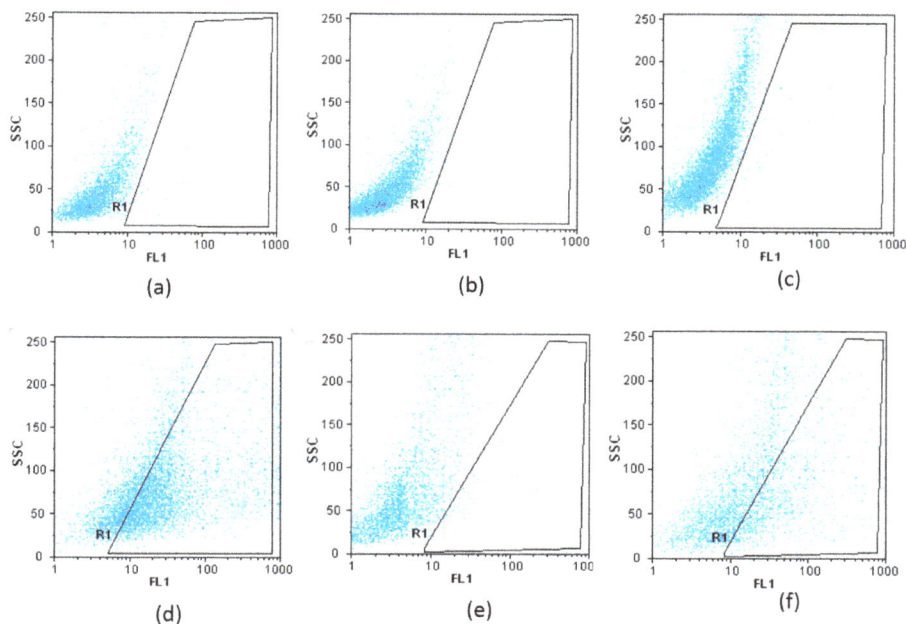

Figure 6. Flow cytometry histograms. (**a,b**) Epithelial cell line (HEp-2); (**c,d**) Epithelial cell line HCT 116 and (**e,f**) Epithelial-like cell line HT 29, before and after binding with aptamer are demonstrated. Experiments were repeated three times.

Table 1. Results of binding between HEp-2, HCT 116 and HT 29 and modified KCHA10a aptamer.

Cell Line	%Gated (R1 Region) Without Aptamer	Aptamer Attached
HEp-2	0.2	0.2
HCT 116	0.43	60.19
HT 29	0.40	29.62

4.3. Electrochemical Experiments

4.3.1. Functionalization of Electrodes

Electrochemical techniques such as cyclic voltammetry offer a great tool to study the self-assembled monolayer of different materials involved in creating linker. In Figure 7, the voltammograms of bare Au electrode, the electrode coated with 11 MUA/EDC/NHS, and the functionalized electrode in the presence of aptamer after BSA treatment, as shown with various colors—green, red and purple—respectively. In these experiments, Ag/AgCl is used as reference electrode.

195

The scan rate was 50 mV/s in the range of −0.2 to 0.8 V. Figure 7 verifies that the amount of current is decreased after every coating step. In other words, the functionalization of the electrode is verified by these voltammograms.

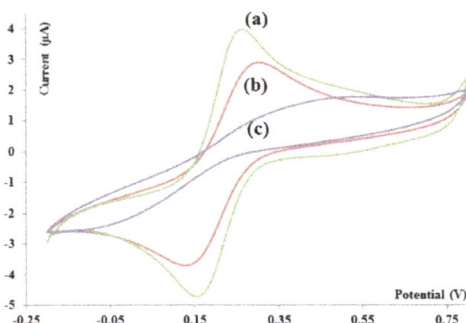

Figure 7. Voltammogram of bare (**a**) Au; (**b**) Au coated with 11MUA/EDC/NHS and (**c**) Au coated with 11MUA/EDC/NHS/APT/BSA in $K_3[Fe(CN)_6]$ 1 mM in PBS (pH 7.4) *vs.* Ag/AgCl.

4.3.2. Elctrochemical Biosensor

The CV results of apta-sensor in the presence of different concentrations of HCT 116 (6, 12, 25, 50, 100, 1000, 17,000 cells/mL) is showed in Figure 8a. Also the effect of different concentrations of HEp-2 cell line as a control cell (6, 12, 100, 1000 cells/mL) on apta-sensor is shown in Figure 8b. As described in Section 5, the CV results confirm the high affinity of colon cancer with aptamer in contrast with the control cells exposed to apta-sensor.

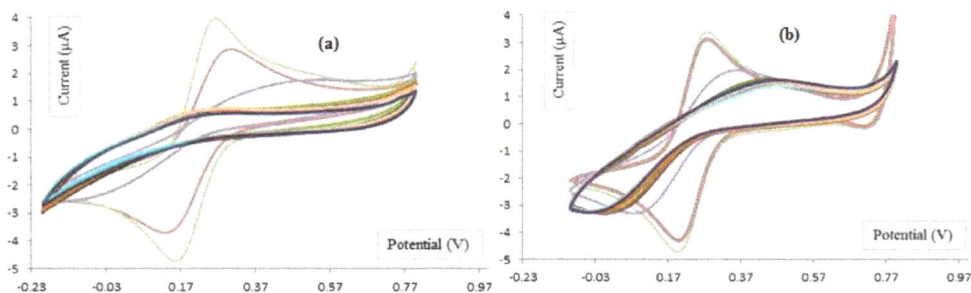

Figure 8. Electrochemical CV results, Au electrode, Au electrode coated with 11-MUA, Au electrode coated with 11-MUA terminated with aptamer and the functionalized electrode conjugated with different concentrations (6, 12, 100, 1000 cell/mL) of (**a**) HCT116 cells; (**b**) HEp-2 cells. Electrolyte was $K_3[Fe(CN)_6]$ 1 mM in PBS (pH 7.4) *vs.* Ag/AgCl.

196

5. Discussion

5.1. Self-Assembly of Monolayers on Electrodes

These experiments performed on the electrodes coated with 11MUA/EDC/NHS and 11MUA/EDC/NHS/APT/BSA. In other words, as the negative charge density on the surface of electrode was increased, the current of $Fe(CN)_6^{3-}$ was decreased accordingly. Therefore, the results shown in Figure 6 demonstrate assembly of monolayers on the surface of Au electrodes.

5.2. Colon Cancer Cell Detection

As shown in Figure 8, the current peak in the presence of cancer and control cells is lower and higher than 1 μA, respectively. This current change may be caused by the higher affinity of cancer cells to conjugate with aptamer, while there is no affinity between the control cells and aptamer. As shown in Figure 8, with the higher concentration of target cells, the higher conjunction between the cells and functionalized electrodes occurs. Consequently, the peak current in each step is decreased. Let us assume the surface of the bare electrode is A_0. It is expected that reduction and oxidation of $Fe(CN)_6$ ions occurs during the CV experiment. When 11-MUA is deposited on the electrode, the surface is equal to A_1. The surface is changed to A_2 after covalent binding of aptamer and linker. Depending on the concentration of cells, the adherent of cells to the surface of electrode changes the surface to A_3. As the results of the change of surface area ($A_0 > A_1 > A_2 > A_3$), the current peak at analyte oxidation (Ipa) and reduction (Ipc) changes as well. In the other words, $Ipa_3 < Ipa_2 < Ipa_1 < Ipa_0$ and $Ipc_0 < Ipc_1 < Ipc_2 < Ipc_3$. The decrease in current oxidation peak (Ipa) and reduction (Ipc), exhibited lower oxidation and reduction of ions because of the changing charge density on the surface of each electrode. The CV results associated with colon cancer models and control cells with different concentrations can be extracted from Figure 8 and shown in Figure 9. This figure shows the difference between the voltammograms of the electrochemical sensor in the presence of HEp2, HT29 and HCT 116 cells.

5.3. Sensitivity of the Aptasensor

The current peaks as a function of voltage for different cell types and cell concentrations are depicted in Figure 10. Based on the results, the saturation threshold of this sensor is 100 cells. When the number of cells exceeds this amount, the sensor loses its sensitivity. Based on the exploration shown in Figure 10, the sensor demonstrates a linear relation between the output current in the range of 1 to 100 cells. Theoretically, at the level of a single cell, the output current of the sensor results in a 0.03 μA change. Also, based on this discussion, the noise level can reach 0.2166 µA and the sensor might be capable of detecting more than seven cells (>0.2166/0.03~7).

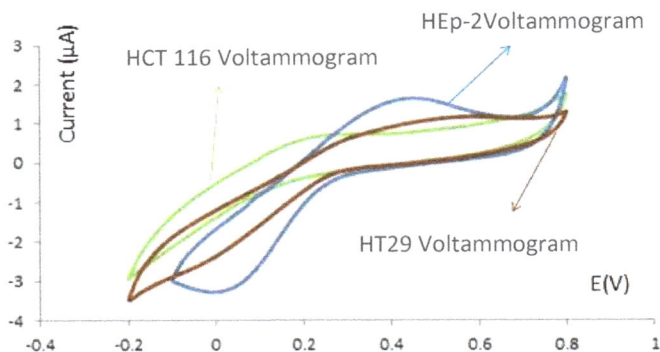

Figure 9. The voltammogram of HEp-2, HT29 and HCT 116 attached to sensor surface in $K_3[Fe(CN)_6]$ 1 mM in PBS (pH 7.4) *vs.* Ag/AgCl.

5.4. Selectivity of the Aptasensor

A comparison between the results of the cancer and control cells illustrates that the sensor function is totally selective for the cancer model cells (Figure 10). The control cells generate a static output because these cells does not increase the output current of sensor. In the case of the cancer cells, an increase in the number of cells leads to an increase of the sensor's output current.

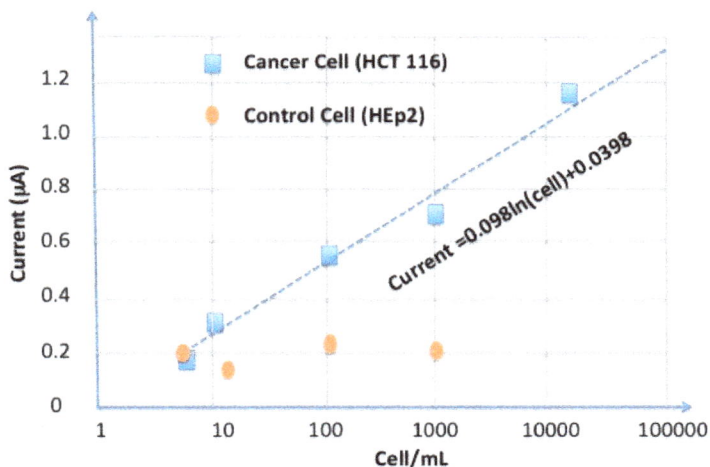

Figure 10. Calibration curve for cancer cells (HCT 116) and control cell (HEp-2).

6. Conclusions

This paper reports the design and implementation of an apta-sensor for colon cancer detection. We have demonstrated and discussed the functionality and applicability of the synthesized KCHA10a aptamer using flow cytometry, fluorescence microscopy and electrochemical experiments. The HCT 116, HT 29 and HEp-2 cell lines were used as colon cancer model and control cells, respectively. The surface of an Au electrode was coated—with SH groups using 11-MUA, EDC/NHS and aptamer. Furthermore, we put forward an apta-sensor demonstrating a linear relation between the numbers of cells (<100) with seven cell resolution. Based on the experimental results and discussions in this paper, the proposed apta-sensor offers high sensitivity and can be a good candidate for colon cancer diagnostics.

Acknowledgments: The authors are grateful for the financial support which provided by The Center for International Scientific Studies and Collaboration (CISSC), Ministry of Science, Research and Technology, Islamic Republic of Iran. The authors would also like to acknowledge the support of NSERC Canada.

Author Contributions: All steps of this research project were performed by Mojgan Ahmadzadeh Raji (PhD candidate) under supervision of Ebrahim Ghafar-zadeh and Ghasem Amoabediny. The consultants of this project are Parviz Tajik and Morteza Hosseini.

Conflicts of Interest: The authors declare no conflict of interest.

References

1. Fast Facts on Colorectal Cancer (CRC). Available online: http://coloncancercanada.ca/fast-facts-on-colorectal-cancer-crc/ (accessed on 1 May 2015).
2. 2014 Canadian Cancer Statistics. Available online: http://www.colorectal-cancer.ca/en/just-the-facts/colorectal/ (accessed on 1 May 2015).
3. Dassie, J.P.; Hernandez, L.I.; Thomas, G.S.; Long, M.E.; Rockey, W.M.; Howell, C.A.; Chen, Y.; Hernandez, F.J.; Liu, X.Y.; Wilson, M.E.; *et al.* Targeted inhibition of prostate cancer metastases with an RNA aptamer to prostate-specific membrane antigen. *Mol. Ther.* **2014**, *22*, 1910–1922.
4. Cao, H.; Ye, D.; Zhao, Q.; Luo, J.; Zhang, S.; Kong, J. A novel apta-sensor based on MUC-1 conjugated CNSs for ultrasensitive detection of tumor cells. *Analyst* **2014**, *139*, 4917–4923.
5. Pourhoseingholi, M.A.; Zali, M.R. Colorectal cancer screening: Time for action in Iran. *World J. Gastrointest. Oncol.* **2012**, *4*, 82–83.
6. Chenga, A.K.H.; Sena, D.; Yua, H.Z. Design and testing of aptamer-based electrochemical biosensors for proteins and small molecules. *Bioelectrochemistry* **2009**, *77*, 1–12.
7. Mascinis, M.; Tombelli, S. Biosensors for biomarkers in medical diagnostics. *Biomarkers* **2008**, *13*, 637–657.
8. Chang, Y.M.; Donovan, M.J.; Tan, W. Using Aptamers for Cancer Biomarker Discovery. *J. Nucleic Acids* **2013**, *2013*, 1–7.

9. Hammarstro, S. The carcinoembryonic antigen (CEA) family: Structures, suggested functions and expression in normal and malignant tissues. *Semin. Cancer Biol.* **1999**, *9*, 67–81.

10. Obernik, B. CEA adhesion molecules: Multifunctional proteins with signal-regulatory properties. *Curr. Opin. Cell Biol.* **1997**, *9*, 616–626.

11. Sadava, D.E. *Cell Biology-Organelle Structure and Function*, 1st ed.; Jones and Bartlett: Charlestown, MA, USA, 1993; Chapter 1; p. 22.

12. Sefah, K.; Meng, L.; Lopez-Colon, D.; Jimenez, E.; Liu, C.H.; Tan, W. DNA Aptamers as Molecular Probes for Colorectal Cancer Study. *PLoS ONE* **2010**, *5*, 1–14.

13. Reth, M. Matching cellular dimensions with molecular sizes. *Nat. Immunol.* **2013**, *14*, 765–767.

14. Stoltenburg, R.; Reinemann, C.; Strehlitz, B. SELEX—A (r) evolutionary method to generate high-affinity nucleic acid ligands. *Biomol. Eng.* **2007**, *24*, 381–403.

15. Medley, C.D.; Smith, J.E.; Tang, Z.; Wu, Y.; Bamrungsap, S.; Tan, W. Gold Nanoparticle-Based Colorimetric Assay for the Direct Detection of Cancerous Cells. *Anal. Chem.* **2008**, *80*, 1067–1072.

16. Tuerk, C.; Gold, L. Systematic evolution of ligands by exponential enrichment: RNA ligands to bacteriophage T4 DNA polymerase. *Science* **1990**, *249*, 505–510.

17. Gopinath, S.C. Methods developed for SELEX. *Anal. Bioanal. Chem.* **2007**, *387*, 171–182.

18. Chiu, T.C.; Huang, C.C. Aptamer-Functionalized Nano-Biosensors. *Sensors* **2009**, *9*, 10356–10388.

19. Liu, Y.; Tuleouva, N.; Ramanculov, E.; Revzin, A. Aptamer-Based Electrochemical Biosensor for Interferon Gamma Detection. *Anal. Chem.* **2010**, *82*, 8131–8136.

20. Bang, G.S.; Cho, S.; Kim, B.G. A novel electrochemical detection method for aptamer biosensors. *Biosens. Bioelectron.* **2005**, *21*, 863–870.

21. Chandra, P.; Singh, J.; Singh, A.; Srivastava, A.; Goyal, R.N.; Shim, Y.B. Gold Nanoparticles and Nanocomposites in Clinical Diagnostics Using Electrochemical Methods. *J. Nanopart.* **2013**, *2013*.

22. Ocaña, C.; Pacios, M.; del Valle, M. A Reusable Impedimetric Aptasensor for Detection of Thrombin Employing a Graphite-Epoxy Composite Electrode. *Sensors* **2012**, *12*, 3037–3048.

23. Balamurugan, S.; Obubuafo, A.; Soper, S.A.; Spivak, D.A. Surface immobilization methods for aptamer diagnostic applications. *Anal. Bioanal. Chem.* **2008**, *390*, 1009–1021.

24. Balamurugan, S.; Obubuafo, A.; McCarley, R.L.; Soper, S.A.; Spivak, D.A. Effect of Linker Structure on Surface Density of Aptamer Monolayers and Their Corresponding Protein Binding Efficiency. *Anal. Chem.* **2008**, *80*, 9630–9634.

25. KineFold Simulation Results. Available online: http://kinefold.curie.fr/cgi-bin/neorequest.pl?batch=0&sim=2-&base=KCHA10a-49060 (accessed on 30 April 2015).

26. Hermanson, G.T. *Bioconjugate Techniques*, 2nd ed.; Academic Press, Inc. of Elsevier: New York, NY, USA, 2008.

27. Freshney, R.I. *Culture of Animal Cells a Manual of Basic Technique*, 5th ed.; Wiley: New York City, NJ, USA, 2005; pp. 199–215.
28. Sheng, W.; Chen, T.; Kamath, R.; Xiong, X.; Tan, W.; Fan, Z.H. Aptamer-Enabled Efficient Isolation of Cancer Cells from Whole Blood Using a Micro fluidic Device. *Anal. Chem.* **2012**, *84*, 4199–4206.

MDPI AG

Klybeckstrasse 64

4057 Basel, Switzerland

Tel. +41 61 683 77 34

Fax +41 61 302 89 18

http://www.mdpi.com/

Sensors Editorial Office

E-mail: Sensors@mdpi.com

http://www.mdpi.com/journal/sensors

www.ingramcontent.com/pod-product-compliance
Lightning Source LLC
Chambersburg PA
CBHW051921190326
41458CB00026B/6365